Enhancement in Surface Engineering Treatments

Enhancement in Surface Engineering Treatments

Edited by **Guy Lennon**

New York

Published by NY Research Press,
23 West, 55th Street, Suite 816,
New York, NY 10019, USA
www.nyresearchpress.com

Enhancement in Surface Engineering Treatments
Edited by Guy Lennon

International Standard Book Number: 978-1-63238-180-4 (Hardback)

Printed in the United States of America.

Contents

Permissions

List of Contributors

Preface

This book has been an outcome of determined endeavour from a group of educationists in the field. The primary objective was to involve a broad spectrum of professionals from diverse cultural background involved in the field for developing new researches. The book not only targets students but also scholars pursuing higher research for further enhancement of the theoretical and practical applications of the subject.

Surface engineering is the phenomenon which facilitates the application of various technologies in a broad spectrum of industrial activities. It was developed by investigating multiple phenomena which result in the destruction of surfaces like abrasion, corrosion, fatigue, and disruption. Soon enough, it was acknowledged that most of the advances in technology were limited with surface requirements. From oil and gas exploitation, to mining and manufacturing, various fields have generated critical problems in technological development due to surfaces. With time, there has been development of new techniques which have served as avenues for enhancing surface properties like abrasion, wear rate and corrosion resistance. Some of the latest advancements in surface engineering have been discussed in this book.

It was an honour to edit such a profound book and also a challenging task to compile and examine all the relevant data for accuracy and originality. I wish to acknowledge the efforts of the contributors for submitting such brilliant and diverse chapters in the field and for endlessly working for the completion of the book. Last, but not the least; I thank my family for being a constant source of support in all my research endeavours.

Editor

Zinc Oxide — Linen Fibrous Composites: Morphological, Structural, Chemical, Humidity Adsorptive and Thermal Barrier Attributes

Narcisa Vrinceanu, Alina Brindusa Petre,
Claudia Mihaela Hristodor, Eveline Popovici,
Aurel Pui, Diana Coman and Diana Tanasa

Additional information is available at the end of the chapter

1. Introduction

The augmented requirement for fibrous supports (yarns) possessing multifunctionality implies powerful emerging multidisciplinary approaches as well as the connection with the traditional scientific disciplines [1]. Finishing processes through nanoparticles were among the first commercial application in textiles domain.

Due to poor fixing of these nanoparticles on the textile surface, these finishes were not resistant to washing. Nanofinishings with improved bonding properties in fabrics and also impart desired wettability will result by using hydrophobic/hydrophilic functional polymer fibrous matrices as dispersion medium for nanoparticles.

Nanoparticles are extremely reactive, due to their high surface energy, and most systems undergo aggregation without protection of their surfaces. To eliminate or minimize generated waste and implement sustainable processes, recently green chemistry and chemical processes have been emphasized for the preparation of nanoparticles [2]. Much attention is now being focused on polysaccharides used as the protecting agents of nanoparticles. As stabilizing agent soluble starch has been selected and as the reducing agent in aqueous solution of $AgNO_3$ for silver nanoparticle growth, and a-D-glucose, has been elected. To maintain noble metal (platinum, palladium and silver) nanoparticles in colloid suspension, Arabinogalactan has been used as a novel protecting agent for [3]. Synthesized platinum, palladium and silver nanoparticles with narrow size distribution have been achieved by using porous cellulose fibers as the stabilizer [4]. Pt nanoparticles can catalyze the carbonization of cellulose and mesoporous amorphous carbon is fabricated in high yields. The results are carbon-based functional composites with metal nanoparticles, showing that self-supporting macroporous sponges of silver, gold and copper oxide, as well as composites of silver/copper oxide or silver/titania can be routinely prepared by heating metal–salt-containing pastes of dextran, chosen as a soft template [5,6]. Polysaccharides could be used as stabilizer to synthesize nanoparticles of metal oxide and sulfides. Zinc oxide nanoparticles can be synthesized using water as a solvent and soluble starch as a stabilizer [7-9] while CdS nanoparticles have been prepared in a sago starch matrix.

In an earlier study, ZnO nanoparticles synthesis can be made with the assistance of MCT-β-CD (monochlorotriazinyl–β -cyclodextrin) by using a sol-gel method [10]. MCT-β-CD, a commercially available β- cyclodextrin with a reactive monochlorotriazinyl group, is used as a stabilizer. The so called anchor group reacting with cellulose hydroxyl radicals and cyclo-dextrin molecule is covalently bonded, to the fiber surface. The stable bound of cyclodextrin onto the textile fibers allows its properties to become intrinsic to the modified supports, thus a new generation of *intelligent textiles possessing* enhanced sorption abilities/capacities, as well as possessing active molecules release wasborn. Besides, as polysaccharide, MCT-β-CD shows interesting dynamic supramolecular associations facilitated both by inter- and intra-molecular hydrogen bonding, and polar groups. When a material is exposed to environmental water vapors, the water molecules firstly reacts with surface polar groups, forming a molecular monolayer.

Zinc oxide (ZnO), an n-type semiconductor, is a very interesting multifunctional material and has promising applications in solar cells, sensors, displays, gas sensors, varistors, piezoelectric devices, electro-acoustic transducers, photodiodes and UV light emitting devices. The adhesion between the ZnO nanoparticles and polymer through simple wet chemical method is rather poor and the nanoparticles may be removed from the host easily. In light of this, it is believed that the hydrothermal method can be a more promising way for fabricating nano-materials because it can be used to obtain products with modified morphological and chemical attributes with high purity, as well as stability in terms of water vapour sorption-desorption. Zn^{2+} ions can penetrate into the interior of linen fibrous support (fabric) easily when soluble salt such as zinc acetate (Zn $(OAC)_2$) is used. Reaction of Zn^{2+} ions leads to crystallization of ZnO nanoparticles within the linen fabric and to the formation of an encapsulated complex in the hydrothermal environment. The formation procedure can be described through two steps as shown in Fig. 1. Firstly, coordination compounds are formed through chelation between Zn^{2+} ions and the hydroxyl groups of linen fabric. Secondly, the in-situ crystallization of Zn chelate complex occurs under the hydrothermal treatment and forms a ZnO- coated linen fabric. The ZnO nanoparticles can thus be attached firmly within the linen fiber surface.

2. Advances in ZnO synthesis

The idea of the interaction of materials with water vapors is a new area of research. Almost all materials have some interaction with moisture that is present in their surroundings. The effects of water can be both harmful and beneficial depending on the material and how it is used. Consequently, the point to determine a correlation between morphological, structural and chemical characterization and the water vapor sorption behavior of the analyzed samples has been emphasized. Consequently, the obtained textiles should find their applicability in textile processing industry subdomains, where a certain level of hydrophilicity/hydrophobicity is mandatory.

The main cause of polymeric materials degradation is the exposure to various factors such as: heat, UV light, irradiation ozone, mechanical stress and microbes. Degradation is promoted by oxygen, humidity and strain, and results in such flaws as brittleness, cracking, and fading [11-13]. There have been research reports targeting nanosized magnetic materials synthesis, having significant potential for many applications.

The applications of ZnO particles are numerous: varistors and other functional devices, reinforcement phase, wear resistant and anti-sliding phase in composites due to their high elastic modulus and strength. Otherwise, ZnO particles exist in anti-electrostatic or conductive phase due to their current characteristics. Few studies have been concerned with the application of ZnO nanoparticles in coatings system with multi-properties. The nano-coatings can be obtained by the traditional coatings technology, i.e., by filling with nanometer-scale materialsBy filling with nano-materials, both structure and functional properties of coatings can be modified. Super-hardness, wear resistant, heat resistance, corrosion resistance, and about function, anti-electrostatic, antibacterial, anti-UV and infrared radiation all or several of them can be realized.

Another idea this paper review was centered to was to study the thermal degradation behavior of some textile nanocomposites made of nano/micron particle grade zinc oxide and linen fibrous supports, and to discuss the thermal degradation mechanism of the above mentioned structures. There is also potential to highlight the effect of the functionalization agent - MCT-β-CD (monochlorotriazinyl–β–cyclodextrin) on the thermal stability and degradation mechanism of ZnO nanocoated linen fibrous samples.

In order to characterize the surface morphology and chemical composition of the treated supports, instrumental methods were conducted to measure the particle sizes of the reduced zinc oxide particles. The understanding of the thermal behavior of these fibers is very important since in general several conventional techniques used in textile processing industry, are conducted at high temperature.

The MCT-β-CD (monochlorotriazinyl–βeta-cyclodextrin) under the trade name CAVATEX or CAVASOL® W7 MCT (CAVATEX) from Wacker Chemie AG, $Zn(OAc)_2$, with an assay of 97%, urea and acetic acid (assay 99%) from CHIMOPAR, cetyltrimethylammonium bromide (CTAB) from Merck Company, with an assay of 97% were utilized

Two 100 % twill linen desized, scoured and bleached supports, each of size 3 cm × 3 cm were used as fibrous support. One of the supports has been coated with a certain concentration of MCT-β-CD (monochlorotriazinyl– β -cyclodextrin) [14-17].

Sample	Specifications
Reference 1	linen fibrous support
Reference 2	ZnO powder hydrothermally synthesized, non-calcinated
Sample 1	Functionalization of linen support with MCT- β –CD (**M**ono**C**hloro**T**riazinyl–**β** -**C**yclo**D**extrin) by exhaustion and thermal treatment
Sample 3	ZnO powder hydrothermally synthesized onto linen fibrous support
Sample 4	ZnO powder hydrothermally synthesized onto functionalized linen fibrous support
Sample 5	ZnO powder hydrothermally synthesized onto functionalized linen fibrous with the assistance of CTAB (**C**etyl **T**rimethyl**A**mmonium **B**romide)
Sample 6	ZnO powder hydrothermally synthesized onto functionalized linen fibrous with the assistance of P123 (**M**ono**C**hloro**T**riazinyl-**β** – **C**yclo**D**extrin)
Sample 7	MCT- β –CD (**M**ono**C**hloro**T**riazinyl–**β** -**C**yclo**D**extrin)

Table 1. Synthesis conditions for each of the sample

2.1. Fundamental technique for synthesizing and characterizing nano-ZnO particles

The review was focused onto the fibrous supports (yarns) previously grafted/functionalized with MCT-β-CD. The grafting process of the textile fabric was performed following two other processes: the exhaustion and squeezing treatment and the heat treatment at 160°C. The purpose of these two treatments was the grafting the linen [18-20].

ZnO nanoparticles were synthesized *in-situ* on linen fibrous supports (yarns) having a certain concentration of MCT-β-CD by using the hydrothermal method. The linen samples with sizes of 30 x30 cm^2 were immersed in the solution prepared as follows: zinc acetate $Zn(CH_3COO)_2$. $2H_2O$, purity – 99%) (0,005 mol/1000mL) as precursor was solved in de-ionized water to form a uniform solution by stirring and then 0,1 mol of urea solution was added drop-wise with constant stirring. Second, the pH value of the mixed solution was adjusted to 5 by adding acetic acid drop wise. The final reaction mixture was then vigorously stirred for two hours at room temperature and poured into 100 mL stainless-steel autoclaves made of Teflon (poly[tetra-fluoroethylene), followed by immersion of the fibrous supports (yarns). Then the autoclaves were placed in the oven for the hydrothermal treatment at 90°C overnight. The autoclaves were then cooled down to room temperature. The treated fabrics were then removed from the autoclaves. The treated fabrics were washed several times with distilled water. After complete washing the composites were dried at 60∘C overnight for complete conversion of the remaining zinc hydroxide to zinc oxide

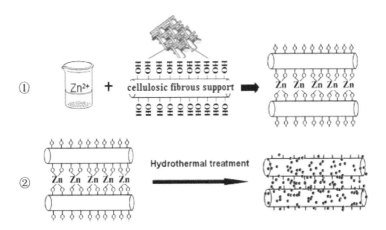

Figure 1. Flow chart for the preparation of nanoparticle coated linen support [10]

Thermal treatment relied into two main stages, into the calcination oven. Firstly, the samples were subjected to an increasing of temperature up to 150°C; secondly the probes were heated up to 350°, 450° respectively.

Scanning Electron Microscope (SEM) images were acquired with a Quanta 200 3D Dual Beam type microscope, from FEI Holland, coupled with an EDS analysis system manufactured by EDAX - AMETEK Holland equipped with a SDD type detector (silicon drift detector). Taking into account the sample type, the analyses have been performed, using Low Vacuum working mode, (as in High Vacuum working type). Both for the acquisition of secondary electrons images (SE – secondary electrons) and EDS type elemental chemical analyses, LFD (Large Field Detector) type detector was used, running at a pressure of 60 Pa, and a voltage of 30 kV.

The ZnO–MCT-β-CD treated fabrics were tightly packed into the sample holder. X-ray Diffraction (XRD) data for structural characterization of the various prepared samples of ZnO were collected on an X-ray diffractometer (PW1710) using Cu-Kα radiation (k = 1.54 Å) source (applied voltage 40 kV, current 40 mA). About 0.5 g of the dried particles were deposited as a randomly oriented powder onto a Plexiglass sample container, and the XRD patterns were recorded at 2θ angles between 20° and 80○, with a scan rate of 1.5°/min. Radiation was detected with a proportional detector [21-25].

2.2. Evaluation of crystallinity

The extent of crystallinity (Ic) was estimated by means of Eq. (1), where I_{020} is the intensity of the 020 diffraction peak at 2θ angle close to 22.6°, representing the crystalline region of the material, and I_{am} is the minimum between 200 and 110 peaks at 2θ angle close to 18°, representing the amorphous region of the material in cellulose fibres [26-28]. I_{020} represents both crystalline and amorphous materials while I_{am} represents the amorphous material.

$$I_C = \frac{I_{020} - I_{am}}{I_{020}} x100(\%) \tag{1}$$

A *shape factor* is used in x-ray diffraction to correlate the size of sub-micrometre particles, or crystallites, in a solid to the broadening of a peak in a diffraction pattern. In the Scherrer equation,

$$\tau = \frac{K \bullet \lambda}{\beta \cos\theta}$$

where K is the shape factor, λ is the x-ray wavelength, β is the line broadening at half the maximum intensity (FWHM) in radians, and θ is the Bragg angle [29]. τ is the mean size of the ordered (crystalline) domains, which may be smaller or equal to the grain size. The dimensionless shape factor has a typical value of about 0.9, but varies with the actual shape of the crystallite.

FTIR was used to examine changes in the molecular structures of the samples. Analysis has been recorded on a FTIR JASCO 660+ spectrometer. The analysis of studied samples was performed at 2 cm^{-1} resolution in transmission mode. Typically, 64 scans were signal averaged to reduce spectral noise.

For the studied samples dynamic vapours sorption (DVS) capacity, at 25°C averaging in the domain of relative humidity (RH) 0-90% has been investigated by using an IGAsorp apparatus, a fully automated gravimetric analyzer, supplied by Hiden Analytical, Warrington - UK). It is a standard sorption equipment, which has a sensitive microbalance (resolution 1μg and capacity 200 mg), which continuously registers the weight of the sample in terms of relative humidity change, at a temperature kept constant by means of a thermostatically controlled water bath. The measuring system is controlled by appropriate software.

To study water sorption at atmospheric pressure, a humidified stream of gas is passed over the sample.

The differential scanning calorimetry analysis (DSC) of fibrous supports - ZnO composites were carried out using a NETZSCH DSC 200 F3 MAIA instrument under nitrogen. Initial sample weight was set as 30-50 mg for each operation. The specimen was heated from room temperature to 350°C at a heating rate of 10°C/min.

3. Prominent assessed features of fibrous composites

From the obtained images it was clearly distinguished the hexagonal shape of ZnO agglomerations and the morphology of linen fibres (Fig.2c).

Zinc Oxide — Linen Fibrous Composites: Morphological, Structural, Chemical, Humidity Adsorptive and
Thermal Barrier Attributes

7

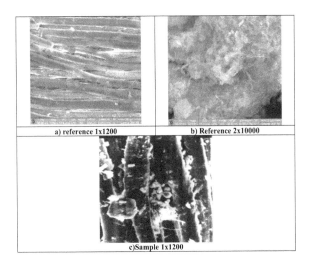

Figure 2. SEM images of: reference samples and of the functionalized linen support with MCT- β –CD sample [10]

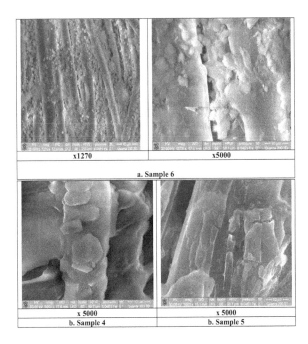

Figure 3. images of some textile composites [10]

The SEM images of functionalized linen supports coated with ZnO with assistance of the studied surfactants (Fig 3 a and b) indicate different shapes of deposited ZnO.

On the other hand, ZnO nanoparticles exhibited hexagonal form like flowers of ZnO nano-crystals, if the treatment was assisted by P123 surfactant (Fig. 3 a) and lamellar morphology if the treatment was assisted by CTAB (Fig. 3 b) respectively.

The particles uniformly cover the fibrous support surface and as results, the fibrous supports surface became coarser after the treatment.

The adhesion strength of ZnO particles on fibrous support is different in terms of the applied surfactant treatment and was tested after repeated washing cycles (1 minute ten times).

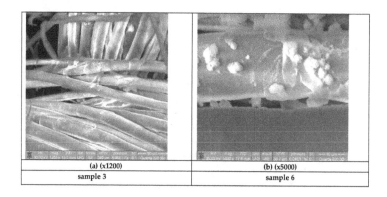

(a) (x1200)	(b) (x5000)
sample 3	sample 6

Figure 4. SEM images of some textile composites after repeated washing cycles [10]

According to the SEM images (Fig 4 a and b), the adhesion strength of ZnO powder hydro-thermally deposited onto functionalized linen fibrous support is superior in the case of functionalized surface (Fig 4 b) compared with the non-functionalized surface (Fig 4a). The functionalization advantage has been evaluated considering the durability of ZnO on the support surface after repeated cycles of washing. After washing the coating particles fell off easily for the ZnO powder hydrothermally synthesized without functionalization, which might have been caused by the weak attaching force.

As shown in Fig. 4, before treatment the diameters of fibrous supports (individual yarns) are about 10 - 20 μm; after treatment SEM image show very clearly the individual yarns, covered by various ZnO aggregates deposition.

3.1. Mechanistic aspect of nanoparticle formation

The shape and the manner of covering depend of linen grafting agent assistance. This result is correlated with the high number of coordinating functional groups (hydroxyl and glucoside groups) of the MCT-β-CD which can form complexes with divalent metal ions [15]. During the synthesis time, it might be possible that the majority of the zinc ions were closely associated with the MCT-β-CD molecules. Based on the previous research, it can be claimed that nucleation and initial crystal growth of ZnO may preferentially occur on MCT-β-CD [16]. Moreover, as polysaccharide, MCT-β-CD showed interesting dynamic supramolecular associations facilitated by inter- and intra-molecular hydrogen bonding, which could act as matrices for nanoparticle growth in size of about 30–40 nm. They aggregated to irregular ZnO–CMC nanoparticles in a further step (Figures 4a) and 4 b). In these figures, SEM images of linen supports coated with ZnO with assistance of the two surfactants show that the nanoparticles exhibited an approximately lamellar morphology and the particles can be seen to be coated on the fibrous support surface (yarn). As result, the fibrous supports (yarns) surface became coarser after the treatment.

In the case of CTAB assistance, on the yarns surfaces large ZnO particles, covering the yarn as a bark are noticeable (Fig. 3b), involving that the large particles may be formed via precipitation followed by a step-like aggregation process. In addition, according to the SEM images of the coated fabric, the uniformity of the fabric coated with ZnO powder hydrothermally synthesized with assistance of CTAB (Cetyl trimethylammonium bromide) is better than that of ZnO powder hydrothermally synthesized in the presence of Pluronic P123 and possesses good washing fastness. The last one has not been measured, but it has *apriori* been evaluated. This phenomenon can be explained by the fact that the repeated cycles of washing and rinsing did not conduct to the washing away of the ZnO particles; subsequently the zinc oxide proven a low extent of washing fastness. This statement is also in a good correlation with the XRD results, claiming a slight shift of ZnO intensity peaks, meaning that the nucleation of the zinc oxide occurred not only the support surface, but also within the nanocavities, due the fibers roughness.

In case of ZnO powder hydrothermally synthesized without any surfactant assistance, the coating particles fell off easily after washing, which might have been caused by the weak attaching force (coordinated bond between ZnO and linen) induced by the deteriorated crystallinity.

The SEM image of functionalized fabric support show very clearly the individual yarns, having diameters of about 10-20 μm, covered by various ZnO aggregates (Fig.4). MCT-β-CD can form complexes with divalent metal ions, due to its high number of coordinating functional groups (hydroxyl and glucoside groups) [31]. There is a possibility that the majority of the zinc ions were closely associated with the MCT-β-CD molecules. Based on the previous research, it can be claimed that nucleation and initial crystal growth of ZnO may preferentially occur on MCT-β-CD [32]. Moreover, as polysaccharide, MCT-β-CD showed interesting dynamic supramolecular associations facilitated by inter- and intra-molecular hydrogen bonding, which could act as matrices for nanoparticle growth in size of about 30–40 nm. They aggregated to irregular ZnO–CMC nanoparticles in a further step.

Element	Wt%	At%
CK	25.34	46.47
NK	15.45	24.30
OK	08.91	12.27
BrL	01.90	00.52
CaK	00.63	00.35
ZnK	47.77	16.10
Matrix	Correction	ZAF

Table 2. Surface composition from EDX measurements at sample 6

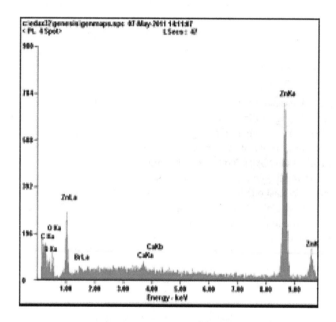

Figure 5. EDX analysis (sample 6) (Wt: weight percent, At: atomic percent). [10]

The outcome of the EDX elemental analysis for sample 6 illustrated in Figure 5 and Table 2, show that surface composition contain approximately 47% ZnO, meaning that ZnO phase represented almost half of the sample mass.

The X-ray diffraction patterns of samples 4-7 compared with reference 2 are represented in Fig. 6:

Figure 6 shows the selected-area diffraction pattern ($2\theta=20\text{-}40^{\circ}$) of the obtained samples. The obtained XRD pattern and indexed lines of ZnO (reference 2) are presented in Figure 6. According to the literature [33], all the diffraction lines are assigned to the wurtzite hexagonal phase structure.

Zinc Oxide — Linen Fibrous Composites: Morphological, Structural, Chemical, Humidity Adsorptive and Thermal Barrier Attributes

11

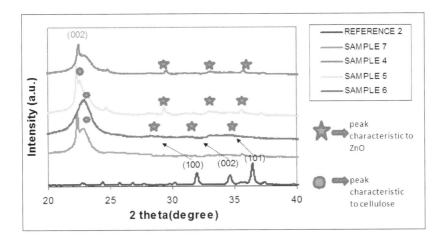

Figure 6. Color online) XRD patterns of: Reference 2; Sample 4; Sample 5; Sample 6; Sample 7[10] The arrows indicate the peaks shift, in terms of working conditions. The height of the peaks has been multiplied by a 40 factor.

The composites patterns (sample 3-6) reveal both the presence of the peaks positions that matched well with those of the ZnO XRD pattern - lines (100), (002) and (101) - and the main peak of cellulose - linen (002) [34]. The small relative intensity of the peaks of the ZnO–linen composites is not well correlated with the EDX analysis, which showed a high content of deposited ZnO. The observed ZnO diffraction lines shift (samples 4 and 5) denotes the fact that the growth of the ZnO takes place not only on the support surface, but also inside the nanocavities due to the fibers roughness.

The intensities of the diffraction peaks decrease when the synthesis takes place with the assistance of the surfactant, that prevent crystal growth in these working conditions (Fig.6).

In Fig. 7, the *FTIR spectrum* of hydrothermally synthesized, non-calcinated ZnO powder exhibited a high intensity broad band at about 430 cm^{-1} due to the stretching of the zinc and oxygen bond.

As shown in the FTIR spectrum of MCT-β-CD, the absorption bands between 1000 and 1200 cm^{-1} were characteristic of the – C –O– stretching on polysaccharide skeleton. A similar band was also observed in synthesized ZnO composites. And two peaks appeared at 1420 and 1610 cm^{-1} corresponding to the symmetrical and asymmetrical stretching vibrations of the carboxylate groups [35]. The peak at 2920 cm^{-1} was ascribed to C–H stretching associated with the ring methane hydrogen atoms. A broad band centered at 3450 cm^{-1} was attributed to a wide distribution of hydrogen-bonded hydroxyl groups. The FTIR spectra indicated that in ZnO–MCT-β-CD nanoparticles, there was the strong interaction, but no obvious formation of covalent bonds between MCT-β-CD) and ZnO.

Water vapors sorption behavior. Isothermal studies can be performed as a function of humidity (0-95%) in the temperature range 5° C to 85° C, with an accuracy of ± 1% for 0 - 90% RH and ±

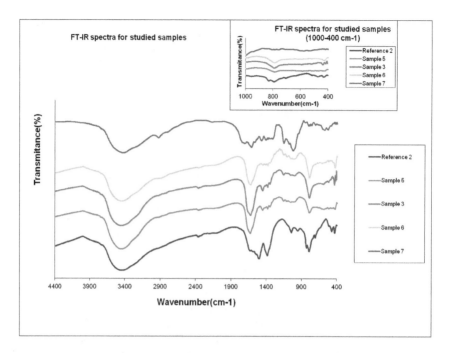

Figure 7. a) FTIR of spectra of: sample 3; sample 5; Sample 6 ; sample 7; Reference 2 [10]

2% for 90 - 95% RH. The relative humidity (RH) is controlled by wet and dry nitrogen flows around the sample. The RH is held constant until equilibrium or until a given time is exceeded, before changing the RH to the next level.

The vapours pressure in the sample room has been achieved by 10 steps of 10% humidity, each of them having a time of equilibrium setting between 10-20 minutes. At each phase, the weight adsorbed by the sample is measured by electromagnetic compensation between tare and sample, when the equilibrium is reached. Apparatus has an anti-condensation system for the cases that vapors pressure is very close/near to that of saturation. The cycle is finished by decreasing in steps of vapors pressure, in order to obtain desorption isotherms, as well.

Prior to measuring of sorption-desorption isotherms, drying of the samples is performed in nitrogen flow (250 mL/min) at 25°C, until the sample weight reached a constant value, at a relative humidity less than 1%.

The sorption/desorption isotherms recorded in these circumstances are shown in Fig.8.

The reference sample (the linen fibrous support – yarn - unfunctionalized) has a smaller sorption capacity compared to that of Sample 6 and Sample 5. High values of water vapors sorption capacity for the two last samples prove the fact that the material surface becomes more hydrophilic, more porous, respectively as it could be observed from hysteresis shape.

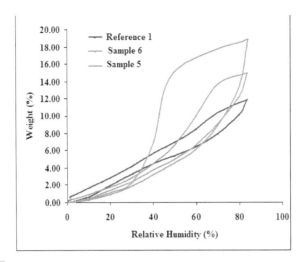

Reference 1; Sample 5; Sample 6 [10]

Figure 8. Comparative plots of rapid isotherms for water vapors sorption for the studied samples:

One of the main objectives of this review was to stress the adsorptive attributes, taking into account the improving of ZnO synthesis conditions. Consequently, the role of P123 in the ZnO synthesis was to obtain a composite with a higher porosity, in order to achieve the surface hydrophilicity, since there is direct correlation between porosity and hydrophilicity [36].

The shape of the moisture sorption isotherms for those two compounds is similar to those characteristic of mesoporous materials (type IV, according to IUPAC classification – with low sorption at low water vapor sorption (adsorption/desorption), moderate sorption at average humidity and rapidly increasing water sorption at high humidity). This type of isotherm describes the sorption behavior of hydrophilic material [37]. When a material is exposed to environmental water vapors, the water molecules firstly react with surface polar groups and form a molecular monolayer.

Based on the sorption studies, the IGAsorp software allows an evaluation of both monolayer and surface area value, by using BET (Brunauer-Emmett-Teller) model (Tabel 2).

Sample	Sorption capacity (%d.b.)	BET analysis	
		A_{BET} (m²/g)	Monolayer (g/g)
Reference 1	11.89	157.010	0.044
Sample 6	14.93	213.99	0.060
Sample 5	18.89	321.39	0.091

Table 3. The main parameters of (water vapors) sorption-desorption isotherms for the studied samples

BET (1) equation is very often used for modeling of the sorption isotherms:

$$W = \frac{W_m \cdot C \cdot RH}{(1-RH) \cdot (1-RH+C \cdot RH)} \tag{2}$$

where:

W- the weight of adsorbed water, Wm- the weight of water forming a monolayer, C – the sorption constant, $p/po=RH$- the relative humidity.

The sorption isotherms described by BET model up to a relative humidity of 40% are in relation to the sorption isotherm and material type. This method is mainly limited for II type isotherms, but can describe the isotherms of I, III and IV type [38-40], as well. The increasing water sorption is reflected both by the augmentation of monolayer and surface area values calculated with BET model (Tabel 3).

In Figure 8 the kinetic curves for humidity (water vapors) sorption/desorption processes for two of the samples are displayed. It is noticed that the time necessary for equilibrium setting for sorption processes is bigger than that of desorption. Sorption rate is smaller than that of desorption.

In Table 4 the dynamic moisture sorption capacity calculation was made using the equation written below, after the samples was kept at RH=90%, until the mass became constant:

Sorption capacity at RH=90% (%) $= \dfrac{W_{RH=90} - W_{RH=0}}{W_{RH=0}} \cdot 100$

As can be observed the obtained values are larger than those in the isotherms, this demonstrates the time necessary for reaching the equilibrium sorption is longer.

Sample	Weight at RH=0% (mg)	Weight at RH=90% (mg)	Sorption dynamic capacity RH=90% (%)	Time (s)	Sorption rate ($\cdot 10^{-3}$ %/s)
Reference 1	4.58	5.14	12.28	32	3.82
Sample 6	5.32	6.34	19.14	50	3.75
Sample 5	5.56	6.83	22.77	40	5.58

Table 4. Water vapor sorption capacity and speed for a longer time (until sample weight remains constant at a relative humidity of 90%)

In case of sample 5, the DVS analysis were made at two temperatures (25 °C and 35 °C), and the influence of this parameter on the sorption/desorption isotherms and kinetics are presented in Figure 8 and Figure 9 respectivelly.

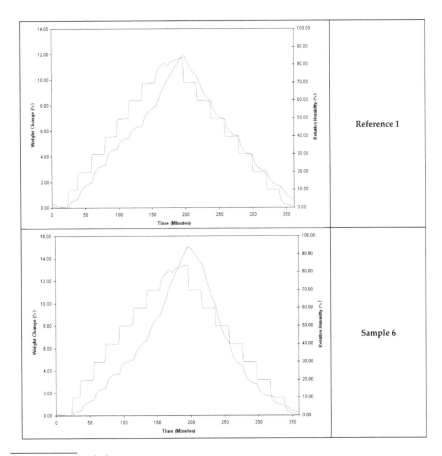

Reference 1; Sample 6 [10]

Figure 9. Kinetic curves for sorption/desorption processes of water vapors in the studied samples

From Table 5, it is noticeable the augmentation of temperature conducts to an increase on vapor sorption capacity of the sample (probably due to the hydrogen bonds formation favoring sorption).

Sample	Sorption capacity (%)
Sample 5_25	18.89
Sample 5_35	31.59

Table 5. Water vapor sorption capacity for sample 5 at both 25 °C and 35 °C respectively

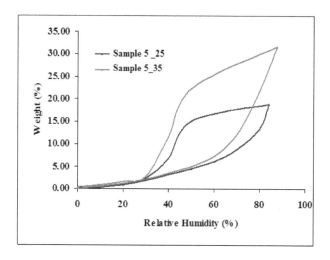

Figure 10. Comparative plots of rapid water vapors sorption/desorption isotherms for sample 5 [10] at both 25 °C and 35 °C respectively

The differences between sorption-desorption speeds of those two temperatures for Sample 5 indexed sample are clearly highlighted by the presence of modified hysteresis and also by the kinetics curves.

3.2. Thermal degradation mechanism of linen fibrous supports treated with ZnO

Considerable attention has been devoted to complete or correlate the results provided by the XRD analysis, with the DSC studies, since the last type of investigation is able to evaluate the crystallization/melting processes.

Vrinceanu et al tested thermal attributes of fibrous supports - ZnO nanocomposites under nitrogen [41] The DSC curves of are shown in figures above.

In the range 370°-395°C, in a typical DSC curve of cellulosic fibres, there is an endothermic peak, which has been shown to be primarily due to the production of laevoglucosan [42].

For linen fibres, this peak is sometimes partly or completely marked by an exothermal effect around 340°C, attributed to a base-catalysed-dehydration reaction that takes place in the presence of alkaline ions, such as those of sodium [43].

From 200 to 250°C a progressive mass loss associated with water release was observed. From the literature it is well known that lignocellulosic fibers degrade in several steps; the cellulose degrades between 310°–360°C, whereas the hemicellulose degrades at about 240°–310°C, and the lignin has been shown to degrade in wide temperature interval (200°–550°C) [44]. Technically speaking, it is not possible to separate the different degradation processes of the fiber components because the reactions are very complex and overlap in the range of 220°–360°C. It is noteworthy that the nanocomposite treated with ZnO nanoparticles with the assistance of

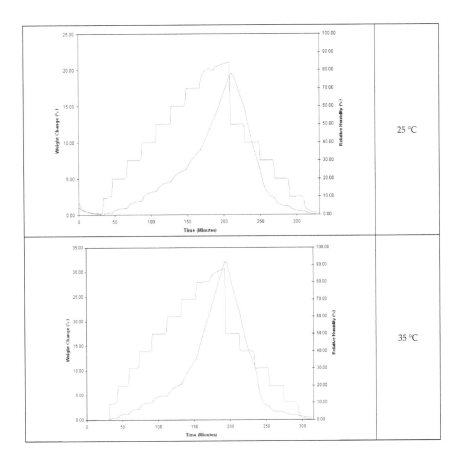

Figure 11. Kinetic curves for sorption/desorption processes of water vapors for sample 5 at both 25 °C and 35 °C respectively [10]

MCT started to decompose at higher temperature than sample treated in the same conditions but without the presence of zinc oxide. Nevertheless, the existence of the MCT on the surface of the probes delayed the thermal degradation of the fibrous linen samples, even the non-treated with the zinc oxide particles.

It can be claimed that cellulose is thermally decomposed through two types of reactions. At lower temperatures, there is a complex process of gradual degradation including dehydration, depolymerisation, oxidation, evolution of carbon monoxide and carbon dioxide, and formation of carbonyl and carboxyl groups, ultimately resulting in a carbonaceous residue forms.

The endothermic band around 260°C from DSC curves (Fig. 14 (a) and (b)) indicates a weight loss. The surface acidity of zinc oxide nanoparticles keeps accelerating the decomposition of

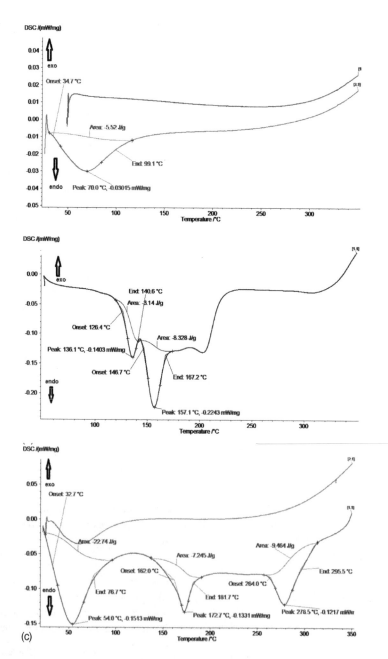

Figure 12. Typical DSC curve under nitrogen for: a Sample 3; b Sample 4; c. Sample 5 [41]

Zinc Oxide — Linen Fibrous Composites: Morphological, Structural, Chemical, Humidity Adsorptive and
Thermal Barrier Attributes

19

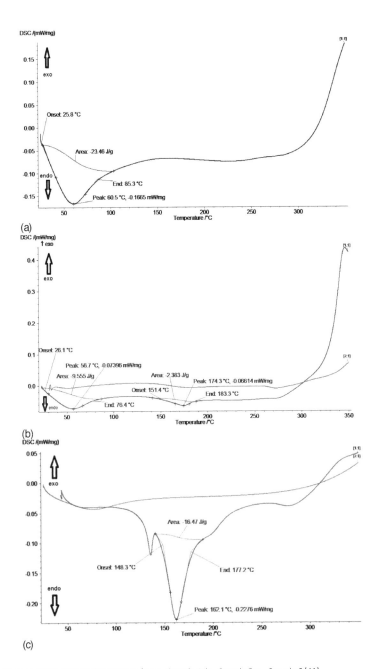

Figure 13. Typical DSC curve under nitrogen for: a – Sample 4; b – Sample 5; c – Sample 6 [41]

the fibrous substrate, as the temperature rises to 310°C, According to the FTIR spectra, a very much lower amount of carbonyl groups is found in the linen - ZnO nanocomposite specimens.

Meanwhile, MCT having a higher thermal conductivity as well as a greater heat capacity value absorbs the heat transmitted from the surroundings and retard the direct thermal impact to the polymer backbone [45,46]. As a consequence, zinc oxide stabilizes the polymer molecules of the underneath substrates and delays the occurrence of major cracking up to 400°C (Fig. 15).

The masking effect of an exothermal reaction on the endothermic cellulose decomposition was clearly highlighted by the behavior of the reference fibrous linen (non-functionalized) subjected to the thermal treatment in N_2; it shows an exothermal peak at 260°C with a decreased enthalpy after the thermal treatment; the exothermal effect is attributable to β-cellulose decomposition as observed in a curve of a cotton sample. Surprisingly, even within the second cycle of thermal treatment, the sample exhibits a similar exothermal peak at 363°C.

4. Summary and outlook

The review has been focused on a series of MCT-β-CD grafted linen fibrous support (yarn) in whose matrices zinc oxide nanoparticles have been introduced with the assistance of two different surfactants. The coating particles fell off easily for the ZnO powder hydrothermally synthesized without any surfactant assistance after washing, which might have been caused by the weak attaching force (coordinated bond between ZnO and linen) induced by the deteriorated crystallinity.

Wetting characteristics are influenced by the type of surfactant used during the hydrothermal synthesis. It is in direct implication onto the relationship between the morphological, structural and chemical attributes and water vapor sorption-desorption behavior. Hydrophilicity of these fibrous composites has increased and based on the sorption/desorption isotherms registered by DSV, BET surface area, as well as XRD measurements were estimated, and assimilated these fibrous composites, set by IUPAC, with mesoporous materials. Humidity loss and drying speed of water from these studied samples depend of the type of surfactant.) A quantification of samples, in terms of their thermal stability has been surveyed, as well. Thus, this paper review intends to develop an innovative and more appropriate synthetic procedure and characterization of nanoscale ZnO coated fibrous composites under favourable conditions, by using the synergic effect of MCT and CTAB/P123 as surfactants. Prominent assessed attributes were emphasized:

• thermal stability and degradation mechanism of ZnO nanocoated linen fibrous samples;

• Cumulative barrier attributes conferred by the new components that interfered in the preparation technique: CTAB/P123 and MCT

These new features are believed to be the promising new lines of exploration of nanoscale ZnO coated fibrous composites in textile area.

Acknowledgements

The authors would like to greatly acknowledge the financial support provided by the two research contracts: /89/1.5/S/49944 POSDRU Project and PN-II-RU-TE-2011-3-0038 project, belonging to "Al.I.Cuza" University of Iasi.

Author details

Narcisa Vrinceanu[1,2], Alina Brindusa Petre[1], Claudia Mihaela Hristodor[1], Eveline Popovici[3], Aurel Pui[1], Diana Coman[2] and Diana Tanasa[1]

1 "Al.I.Cuza" University of Iasi, Iasi, Romania

2 "L.Blaga" University of Sibiu, Romania

3 Al.I.Cuza" University of Iasi, Faculty of Chemistry, Departament of Materials Chemistry, Romania

References

[1] Weber, J., Futterer, C., Gowri, V.S., Attia, R., Viovy, J.L., La Houille Blanche 5:40 (2006); Saxana, M, Gowri. V.S., J Polym Compd 24:428 (2003); Gowri, V.S., Saxena, MJ Chem Technol 14:145 (1997)

[2] Raveendran, P., Fu, J., Wallen, S.L., Completely "green" synthesis and stabilization of metal nanoparticles, J. Am. Chem. Soc. 125, 13940–13941 (2003)

[3] Mucalo, M.R., Bullen, C.R., Arabinogalactan from the Western larch tree: a new, purified and highly water-soluble polysaccharide-based protecting agent for maintaining precious metal nanoparticles in colloidal suspension, J. Mater.Sci. 37, 493–504 (2002)

[4] He, J.H., Kunitake, T., Nakao, A., Facile in situ synthesis of noble metal nanoparticles in porous cellulose fibers, Chem. Mater. 15, 4401–4406 (2003)

[5] He, J.H., Kunitake, T., Nakao, A., Facile fabrication of composites of platinum nanoparticles and amorphous carbon films by catalyzed carbonization of cellulose fibers, Chem. Commun. 4, 410–411 (2004)

[6] Walsh, D., Arcelli, L., Ikoma, T., Tanaka, J., Mann, S., Dextran templating for the synthesis of metallic and metal oxide sponges, Nat. Mater. 2, 386–390 (2003)

[7] Vigneshwaran, N., Kumar, S., Kathe, A.A., Varadarajan, P.V., Prasad, V., Functional finishing of cotton fabrics using zinc oxide–soluble starch nanocomposites, Nanotechnology 17, 5087–5095 (2006)

[8] Ma, X.F., Chang, P.R., Yang, J.W., Yu, J.G.,. Preparation and properties of glycerol plasticized-pea starch/zinc oxide–starch bionanocomposites, Carbohydr. Polym. 75, 472–478 (2009)

[9] Radhakrishnan, T., Georges, M.K., Nair, P.S., 2007. Study of sago starch–CdS nanocomposite films: Fabrication, structure, optical and thermal properties, J.Nanosci. Nanotechnol. 7, 986–993 (2007)

[10] Diana Tanasa, Narcisa Vrinceanu, Alexandra Nistor,Claudia Mihaela Hristodor, Eveline Popovici, Ionut Lucian Bistricianu, Florin Brinza1, Daniela-Lucia Chicet,Diana Coman, Aurel Pui, Ana Maria Grigoriu, Gianina Broasca, Zinc oxide-linen fibrous composites: nmorphological, structural, chemical and humidity adsorptive attributes, Textile Research Journal, 82(8) 832–844 (2012)

[11] Chandramouleeswaran, S., Mhaskel, S., Kathe, A.A., Varadarajan, P.V., Prasad, V., Vigneshwaran, N., Functional behaviour of polypropylene/ZnO–soluble starch nanocomposites, Nanotechnology 18, 385702 (2007)

[12] Tang Z K, Wong G K L, Yu P, Kawasaki M, Ohtomo A, Koinuma H and Segawa Y Appl. Phys. Lett. 72 3270 (1998)

[13] Corrales T, Catalina F, Peinado C, Allen NS, Fontan E. Photooxidative and thermal degradation of polyethylenes interrelationship by chemiluminescence, thermal gravimetric analysis and FTIR data. J Photochem Photobiol. A 2002;147:213-24

[14] Gawas UB, Verenkar VMS, Mojumdar SC. Synthesis and characterization of Co0.8Zn0.2Fe2O4 nanoparticles. J Therm Anal Calorim. 2011;104:879-883.

[15] Gonsalves LR, Mojumdar SC, Verenkar VMS. Synthesis and characterization of Co0.8Zn0.2Fe2O4 nanoparticles. J Therm Anal Calorim. 2011;104:869-873

[16] Gonsalves LR, Mojumdar SC, Verenkar VMS. Synthesis of cobalt nickel ferrite nanoparticles via autocatalytic decomposition of the precursor. J Therm Anal Calorim. 2010;100:789-792.

[17] Gawas UB, Mojumdar SC, Verenkar VMS. Synthesis of cobalt nickel ferrite nanoparticles via autocatalytic decomposition of the precursor. J Therm Anal Calorim. 2010;100:867-871.

[18] Verdu J, Rychly J, Audouin L. Polym Degrad Stab. 2003;79:503-9.

[19] Allen NS, Edge M, Corrales T, Childs A, Liauw CM, Catalina F, et al. Ageing and stabilization of filled polymers: an overview. Polym Degrad Stab. 1998;61:183-99.

[20] Mojumdar SC, Moresoli C, Simon LC. Legge RL. Edible wheat gluten (WG) protein films: Preparation, thermal, mechanical and spectral properties. J Therm Anal Calorim. 2011; 104:929-936.

[21] Gawas UB, Mojumdar SC, Verenkar VMS. J Therm Anal Calorim. 2009;96:49-52

[22] Mocanu AM, Odochian L, Apostolescu N, TG-FTIR study on thermal degradation in air of some new diazoaminoderivatives. J Thermal Anal Calorim. 2010;100 (2): 615-622.

[23] Singhal M, Chhabra V, Kang P, Shah DO. Mater Res Bull. 1997; 32:239-247

[24] Hingorani S, Pillai V, Kumar P, Multani MS, Shah DO. Mater Res Bull. 1993; 28:1303-1310.

[25] Fan Q, John J, Ugbolue SC, Wilson AR, Dar YS, Yang Y. AATCC Rev. 2003; 3(6):25.

[26] Grigoriu A.M., Cercetări în domeniul compusilor de incluziune ai ciclodextrinelor si al derivatilor acestora cu aplicatii în industria textilă. Ph. D. Diss., Iasi (2009)

[27] Reuscher H, Hinsenkorn R. BETA W7 MCT-new ways in surface modification. J Incl Phenom Macrocycl. 1996;25:191–196.

[28] Ogata N, Ogawa T, Ida T, Yanagawa T, Ogihara T, Yamashita A. Sen'i Gakkaishi. 1995;51(9), 439.

[29] Patterson A. The Scherrer Formula for I-Ray Particle Size Determination. Phys Rev. 1939;56 (10): 978–982.

[30] Xiao-Juan J, Pascal Kamdem D. Chemical composition crystallinity and crystallite cellulose size in populus hybrids and aspen. Cellul Chem Technol. 2009;43(7-8): 229-234

[31] Reuscher H., Hinsenkorn R., Journal of inclusion phenomena and macrocyclic chemistry, 25, 191–196 (1996)

[32] Taubert, A., Wegner, G., Formation of uniform and monodisperse zincite crystals in the presence of soluble starch. J. Mater. Chem. 12, 805–807 (2002)

[33] J. Yu et al. Bioresource Technology 100 2832–2841 (2009)

[34] Sunkyu Park1,3, John O Baker1, Michael E Himmel1, Philip A Parilla and David K Johnson, Cellu lose crystallinity index: measurement techniques and their impact on interpreting cellulase performance, Biotechnology for Biofuels 2010, 3:10 Jiugao, Yu, Jingwen, Yang, Baoxiang, Liu, Xiaofei, Ma, Preparation and characterization of glyc-erol plasticized-pea starch/ZnO–carboxymethylcellulose sodium nanocomposites, Bi-oresource Technology 100, 2832–2841 (2009)

[35] Gu, F., Wang, S.F., Lu, M.K., Zhou, G.J., Xu, D., Yuan, D.R. Langmuir, 20: 3528 (2004)

[36] http://test.ttri.org.tw/neweng/RD/images/pub3.pdf

[37] Brunauer, S., Deming, L.S., Deming, W.E., Teller, E. On a Theory of the van der Waals adsorption of gases, J Am Chem Soc, 62(7): 1723-1732 (1940

[38] Guggenheim, E. A. Application of Statistical Mechanics, Clarendon Press, Oxford, 186-206 (1966).

[39] Anderson, R.B. Modifications of the Brunauer, Emmett and Teller Equation, J Am Chem Soc, 68(4): 686-691 (1946)

[40] de Boer, J.H. The Dynamical Character of Adsorption, 2nd ed., Clarendon Press, Oxford, 200-219 (1968).

[41] Vrinceanu , N., Tanasa, D., Hristodor C.M., Brinza, F., Popovici E., Gherca D., Pui A., Coman, D., Carsmariu A., Bistricianu I., Broasca, G., Synthesis and characterization of zinc oxide nanoparticles Application to textiles as thermal barriers, J Therm Anal Calorim, DOI 10.1007/s10973-012-2269-7 (2012)

[42] Ye DY, Farriol X. Preparation and characterization of methylcellulose from Miscanthus Sinensis. Cellul. 2005; 12, 507.

[43] Revola JF, Dietricha ND D, Goring AI. Effect of mercerization on the crystallite size and crystallinity index in cellulose from different sources. Can J Chem. 1987;65, 1724.

[44] [44].Nevell TP, Zeronian S H. Cellulose Chemistry and Its Applications M. Chichester: Ellis Horwood Ltd., 1985;423–454.

[45] Lee HJ, Yeo SY, Jeong SH. Antibacterial Effect of Nanosized Silver Colloidal Solution on Textile Fabrics. J Mater Sci. 2003;38:2199–2204.

[46] Riva A, Algaba IM, Montserrat P. Action of a finishing product in the improvement of the ultraviolet protection provided by cotton fabrics. Modelisation of the e * ect. Cellul. 2006;13:697–704.

Chapter 2

Surface Modification by Friction Based Processes

R. M. Miranda, J. Gandra and P. Vilaça

Additional information is available at the end of the chapter

1. Introduction

The increasing need to modify the surface's properties of full components, or in selected areas, in order to meet with design and functional requirements, has pushed the development of surface engineering which is largely recognised as a very important field for materials and mechanical engineers.

Surface engineering includes a wide range of processes, tailoring chemical and structural properties in a thin surface layer of the substrate, by modifying the existing surface to a depth of 0.001 to 1.0 mm such as: ion implantation, sputtering to weld hardfacings and other cladding processes, producing typically 1 - 20 mm thick coatings, usually for wear and corrosion resistance and repairing damaged parts. Other deposition processes, such as laser alloying or cladding, thermal spraying, cold spraying, liquid deposition methods, anodising, chemical vapour deposition (CVD), and physical vapour deposition (PVD), are also extensively used in surface engineering. Hardening by melting and rapid solidification and surface mechanical deformation allow to change the properties without modifying its composition [1].

Friction based processes comprise two manufacturing technologies and these are: Friction Surfacing (FS) and Friction Stir Processing (FSP). The former was developed in the 40's [2] and was abandoned, at that time, due to the increasing developments observed in competing technologies as thermal spraying, laser and plasma. Specially laser surface technology has largely developed in the following years in hardening, alloying and cladding applications and is now well established in industry. However, FS as a solid state processing technology, was brought back for thermal sensitive materials due to its possibility to transfer material from a consumable rod onto a substrate producing a coating with a good bonding and limited dilution.

The patented concept of Friction Stir Welding in the 90's [5] opened a new field for joining metals, specially light alloys and friction stir processing emerged around this concept.

FSP uses the same basic principles as friction stir welding for superficial or in-volume processing of metallic materials. Applications are found in localized modification and microstructure control in thin surface layers of processed metallic components for specific property enhancement. It has proven to be an effective treatment to achieve major microstructural refinement, densification and homogenisation of the processed zone, as well as, to eliminate defects from casting and forging [6-8]. Processed surfaces have enhanced mechanical properties, such as hardness, tensile strength, fatigue, corrosion and wear resistance. A uniform equiaxial fine grain structure is obtained improving superplastic behaviour. FSP has also been successfully investigated for metal matrix composite manufacturing (MMCs) and functional graded materials (FGMs) opening new possibilities to chemically modify the surfaces [9].

However, FSP has some disadvantages, the major of which is tool degradation and cost, which limits its wider use to high added value applications. Therefore, friction surfacing (FS) emerged again.

This chapter will focus on the mechanisms involved in both FSP and FS and their operating parameters, highlighting existing and envisaged applications in surface engineering, based on the knowledge acquired from ongoing research at the author's institutions.

2. Friction stir processing

2.1. Fundamentals

Friction Stir Processing (FSP) is based on the same principles as friction stir welding (FSW) and represents an important breakthrough in the field of solid state materials processing.

FSP is used for localized modification and microstructural control of surface layers of processed metallic components for specific property enhancement [6]. It is an effective technology for microstructure refinement, densification and homogenisation, as well as for defect removal of cast and forged components as surface cracks and pores. Processed surfaces have shown an improvement of mechanical properties, such as hardness and tensile strength, better fatigue, corrosion and wear resistance. On the other hand, fine microstructures with equiaxed recrystallized grains improve superplastic behaviour of materials processing and this was verified for aluminium alloys [7]. More recently the introduction of powders preplaced on the surface or in machined grooves allowed the modification of the surfaces, producing coatings with characteristics different from the bulk material, or even functionally graded materials to be discussed later in this chapter. The process has still limited industrial applications but is promising due to its low energy consumption and the wide variety of coating / substrate material combinations allowed by the solid state process.

A non-consumable rotating tool consisting of a pin and a shoulder plunges into the workpiece surface. The tool rotation plastically deforms the adjacent material and generates frictional heat both internally, at an atomic level, and between the material surface and the shoulder. Localized heat is produced by dissipation of the internal deformation energy and interfacial friction between the rotating tool and the workpiece. The local temperature of the substrate rises to the range where it has a viscoplastic behaviour beneficial for thermo-mechanical processing. When the proper thermo-mechanical conditions, necessary for material consolidation are achieved, the tool is displaced in a translation movement. As the rotating tool travels along the workpiece, the substrate material flows, confined by the rigid tool and the adjacent cold material, in a closed matrix like forging manufacturing process. The material under the tool is stirred and forged by the pressure exerted by the axial force applied during processing as depicted in Figure 1.

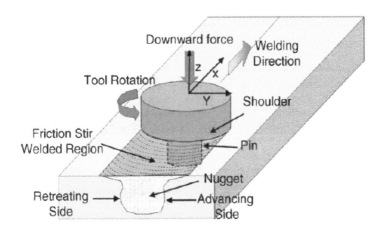

Figure 1. Schematics of friction stir processing.

The material structure is refined by a dynamic recrystallization process triggered by the severe plastic deformation and the localised generated heat. Homogenization of the structure is also observed along with a defect free modified layer of micrometric or nanometric grain structure.

FSP is considered an environmentally friend technology due to its energy efficiency and absence of gases or fumes produced. Table 1 summarizes the major benefits of FSP considering technical, metallurgical, energy and environment aspects.

Technical	Processed depth controlled by the pin length
	One-step technique
	No surface cleaning required
	Good dimensional stability since it is performed under solid state
	Good repeatability
	Facility of automation
Metallurgical	Solid state process
	Minimal distortion of parts
	No chemical effects
	Grain refining and homogenization
	Excellent metallurgical properties
	No cracking
	Possibility to treat thermal sensitive materials
Energy	Low energy consumption since heat is generated by friction and plastic deformation
	Energy efficiency competing with fusion based processes as laser
Environmental	No fumes produced
	Reduced noise
	No solvents required for surface degreasing and cleaning

Table 1. Major benefits of friction stir processing

Analysing the cross section of a friction stir processed surface, three distinct zones can be identified and these are: the nugget or the stirred zone (SZ), the thermomechanically affected zone (TMAZ) and the heat affected zone (HAZ) as shown in Fig. 2.

The nugget, just below the pin and confined by the shoulder width, is the area of interaction where severe plastic deformation occurs. The raise in local temperature due to internal friction and the generated friction between the shoulder and the surface along with the high strain, promotes a dynamically recrystallized zone, resulting in the generation of fine homogeneous equiaxial grains in the stirred zone and precipitate dissolution. Though this is a solid state process, the maximum temperature can be of about 80% of the fusion temperature. Ultrafine-grained microstructures with an average grain size of 100-300 nm in a Mg-Al-Zn alloy were observed in a single pass under cooling [8]. These micro and nano structures are responsible for increases in hardness and wear behaviour reported by several researchers studying different types of alloys under different processing conditions.

The thermo-mechanically affected zone (TMAZ) is immediately adjacent to the previous and, in this zone, the deformation and generated heat are insufficient to generate new grain formation, thus, deformed elongated grains are observed with second phases dispersed in the grain boundaries. Though new grain nucleation may be observed, microstructure remains elongated and deformed. The hardness is higher than in the heat-affected zone due to the high dislocations density and sub-boundaries caused by plastic deformation.

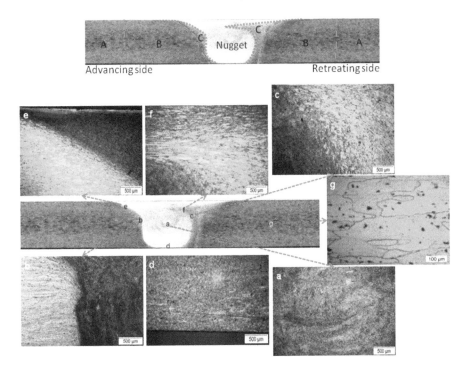

Figure 2. A typical macrograph showing various microstructural zones in FSW of AA2024-T351

In the heat affected zone (HAZ) no plastic deformation is experienced but heat dissipated from the stirred zone into the bulk material can induce phase transformations, depending on the alloys being processed, as precipitate coarsening, localized aging or annealing phenomena.

The non symmetrical character of the process is also evident (Fig. 2). The advancing side is usually referred to the one where the rotating and travel movements have the same direction, while in the retreating side these have opposite directions. On the advancing side, the recrystallized zone is extended and the nugget presents a sharp appearance. The relative velocity between the tool and the base material is higher due to the combination of tool rotation and translation movement. As such, plastic deformation is more intense, thus, the increase in the degree of deformation during FSP, results in a reduction of recrystallized grain size, extending the fine-grain nugget region to the advancing side. Hardness can be higher than in the thermo-mechanically affected zone, but typically lower than in the base material, whenever it is a heat treatable alloy hardenable by aging. The Hall Petch equation establishes a relation between grain size and yield strength and states that these vary in opposite senses [10]. So, in the nugget yield strength is seen to be much higher than in the base material and this is a major result from this process.

2.2. Processing parameters

Operating or processing parameters determine the amount of plastic deformation, generated heat and material flow around the non-consumable tool.

The tool geometry is of major relevance as far as material flow is concerned. Two main elements constitute the tool and these are the pin and the shoulder. Geometrical features such as pin height and shape, shoulder surface pattern and diameter, have a major influence on material flow, heat generation and material transport volume, determining the final microstructure and properties of a processed surface. Several tools have been designed and patented for both FSW and FSP.

The pin (Fig. 3) can be cylindrical or conical, flat faced, threaded or fluted to increase the interface between the probe and the plasticized material, thus intensifying plastic deformation, heat generation and material mixing. The pin length determines the depth of the processed layer. However, since FSP usually aims to produce a thin fine-grained layer across a larger surface area, pinless tools with larger shoulder diameters can also be used.

Figure 3. Example of tool geometries

Shoulder profiles aim to improve friction with the material surface generating the most part of frictional heat involved in the process. Shoulders can be concave, flat or convex, with grooves, ridges, scrolls or concentric circles as depicted in Figure 4.

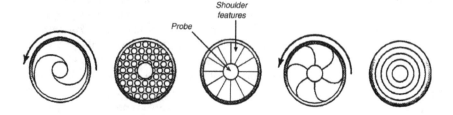

Figure 4. Examples of shoulder geometries

The major processing parameters are the tool rotation and traverse speeds, the axial force and the tilt angle:

• Tool rotation and traverse speeds

These parameters, individually or in combination, affect the plastic deformation imposed onto the material and, thus, the generated heat. An empirically accepted concept divides processing into two main classifications: "cold" and "hot". Cold processing is the one where the ratio between rotating and traverse speed is below 3 rpm/mm and hot processing when this ratio is above 6. Though there is no scientific basis for this border line, it is, however, noticeable that increasing this ratio, the SZ is larger and a very fine structure is observed, while under "cold" conditions the SZ is not well defined since the heat generated is insufficient to promote grain recrystallization. So, increasing the tool rotation speed, plastic deformation is more intense and so is generated heat enabling more material mixing. Therefore, it is possible to achieve a smaller grain size of equiaxial homogeneous grains with precipitate dissolution.

Transverse speed mostly affects the exposure time to frictional heat and material viscosity. Low traverse speeds result in larger exposure times at higher process temperatures.

• Tool axial force

This parameter affects friction between the shoulder and the substrate surface generating and promoting material consolidation. High axial force causes excessive heat and forging pressure, obtaining grain growth and coarsening, while low axial forces lead to poor material consolidation, due to insufficient forging pressure and friction heating. Excessive force may also result in shear lips or flashes with excessive height of the beads on both the advancing and retreating sides, causing metal thinning at the processed area and poor yield and tensile properties. So, surface finishing is much controlled by the axial or forging force.

• Tilt angle

The tilt angle is the angle between the tool axis and the workpiece surface. The setting of a suitable tilting towards the traverse direction assures that the tool moves the material more efficiently from the front to the back of the pin and improves surface finishing.

The effect of the different process parameters has been widely documented by several authors and they are all unanimous that plastic deformation and consequent heat generation are essential to establish the viscoplastic conditions necessary for the material flow and to achieve good consolidation. Thus, a tilt angle of 2-4º is usually used in practice.

Insufficient heating, caused by poor stirring (low tool rotational rates), a high transverse speed or insufficient axial force, results in improper material consolidation with consequent low strength and ductility. Raising heat will cause grain size to decrease to a nanometric scale improving material properties. However, a very significant increase in tool rotation rate, axial force or a very low transverse speed may result in high non desired temperature, slow cooling rate or excessive release of stirred material with property degradation.

2.3. Multiple passes

In order to process large areas in full extent, multiple-passes are used. These can be run separately or overlapped. An overlap ratio (OR) was defined to characterize the overlap between passes and defined by equation 1 [7].

$$OR = 1 - \left[\frac{l}{d_{pin}} \right] \tag{1}$$

Where l is distance between centres of each pass and d_{pin} is the maximum diameter of the pin. From this equation, fully overlapped passes have an OR=1 and OR decreases, when increasing the distance between passes. For an OR<0 no overlap of the nuggets exists.

There are two types of material modification by Friction Stir Processing, the in-volume FSP (VFSP) consisting on the modification of the full thickness of the processed materials and the surface FSP (SFSP) which consists in the surface modification up to depth of about 2 mm.

Figure 5 depicts the effect of OR in two Al alloys, a heat treatable (AA7022-T6) and a non heat treatable one (AA5083-O) with different number of passes and overlap ratios.

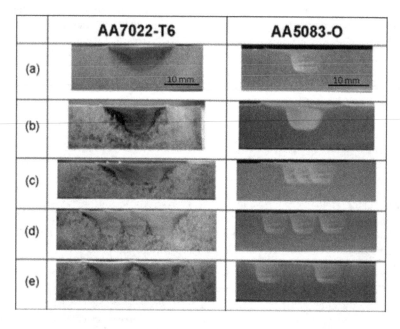

Figure 5. Cross sections of the samples processes with different treatments a) one pass with OR=1; b) four passes with OR=1; c) three passes with OR=1/2; d) three passes with OR=0 and e) two passes with OR=-1 [7]

In this study [7] the authors showed that AA5083-O alloy needed at least three passes in the same location to produce a homogeneous processed area, while the AA7022-T6 alloy only needed one pass, since this is a heat treatable alloy. Grain size reduced from 160 μm (AA7022-T6) and 106 μm (AA5083-O) to an average grain size of about 7.1 and 5.9 μm, respectively. The highest hardness value was located in the nugget due to a significantly decrease in the grain size. This results that in AA7022-T6 alloy the hardness is lower in the nugget than in the base material because it is a heat treatable aluminium alloy and in the AA5083-O alloys the hardness in the nugget is higher than in the base material which is a typical behaviour of non-heat treatable alloys. A significant increase in the formability of the materials was observed due to the increase of the materials ductility resulting from the refinement of the grain size, increasing the maximum bending angle in four times for the SFSP treatment and twelve times for the VFSP treatment in the AA7022-T6 samples. In AA5083-O samples an increase in the maximum bending angle around 1.5 times for the SFSP treatment and about 2.5 times for VFSP treatment was observed.

The overlapping direction in multipass Friction Stir Processing (FSP) was also seen to have a major influence on the surface geometrical features [11]. Structural and mechanical differences were observed in a AA5083-H111 alloy when overlapping by the advancing side (AS) direction or by the retreating side (RS) one. Overlapping by the retreating side was found to generate smoother surfaces, while overlapping by the advancing side led to more uniform thickness layer (Fig. 6). This result is quite relevant from a practical point of view since when the aim of processing large areas in multiple passes procedure is to increase the depth of the processed zone, overlapping of successive passes should be performed by the advancing side of the previous pass. If surface finishing is to be maximised to prevent finishing operations, over-lapping on the previous pass in the retreating side produces very low rough surfaces.

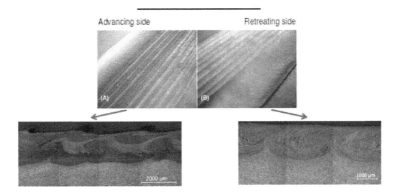

Figure 6. Macro and micrographs of cross sections in friction stir processed surfaces when overlapping by the advancing and by the retreating sides [11]

Hardness within the processed layer increased by 8.5 % and was seen to be approximately constant between passes. The mechanical resistance and toughness under bending were improved by 18 % and 19 %, respectively.

Bending test curves are presented in Figure 7. The processed surfaces were tested under tensile and compression loads. Different behaviours were observed for each bending specimen. Surface modification by multi-pass FSP resulted in an increase of the maximum load supported for all samples and up to a maximum of 18 % for the compression solicitation of the sample produced when overlapping by the RS (Fig. 8). FSP produced a thin layer of a fine equiaxial recrystallized grain structure and homogeneous precipitation dispersion, enhancing material strength.

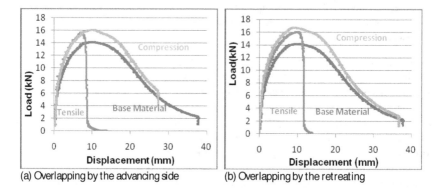

(a) Overlapping by the advancing side   (b) Overlapping by the retreating

Figure 7. Load vs. displacement plot of the bending tests of the FSP samples produced when overlapping by (a) AS and (b) RS [11].

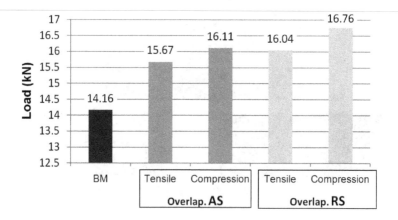

Figure 8. Maximum load attained by FSP samples under different test conditions relatively to the base material [11].

2.4. Applications and performance

Friction stir processing can be used to locally refine microstructures and eliminate casting defects in selected locations, where property improvements can enhance component perform-ance and service lifetime. For instances, aluminium castings contain porosities, segregated phases and inhomogeneous microstructures which contribute to property degradation. Microstructural casting defects, such as: coarse precipitates and porosities increase the possibility of rupture due to the intragranular nucleation of micro-cracks during material deformation. Precipitates are less capable of plastic deformation than the matrix, so cavity nucleation is very frequent, whether caused by a disconnection from the matrix or the rupture of precipitates.

Friction stir processing allows the breakage of large precipitates and their dispersion in a homogeneous matrix, increasing the material capability to withstand deformation, since it results in a higher level of crack closure. Additionally, mechanical properties such as ductility, fatigue strength and formability, are improved.

On the other hand, a large number of small precipitates increases the material resistance to deformation and hence its strength, as they act as barriers or anchorage points to dislocations movements. A uniform equiaxial fine grain structure is also essential to enhance material superplastic behaviour. Friction stir processing generates fine microstructure and equiaxed recrystallized grains which leads either to an increase in strain rate or a decrease in the temperature at which superplasticity is achieved.

In the FSP of an aluminium cast alloy ADC12, Nakata et al. [12] applied multiple-passes to increase tensile strength to about 1.7 times that of the base material. The hardness profile of processed layer was uniform and about 20 HV higher than that of the cast material. The observed increase in tensile strength was attributed to the elimination of the casting defects such as porosities, an homogeneous redistribution of fine Si particles and a significant grain refinement to 2–3 μm. Santella et al. [13] investigated the use of friction stir processing to homogenise hardness distributions in A319 and A356 cast aluminium alloys. Hardness and tensile strength were increased relatively to the cast base material.

Similar results were also reported in the friction stir processing of magnesium based alloys. A.H. Feng and Z.Y. Ma et al. [14] combined FSP with subsequent aging to enhance mechanical properties of Mg-Al-Zn castings

Chang et al. [8] obtained a significant improvement of mechanical properties as the mean hardness measured at the ultrafine-grained zone reached approximately 120HV (twice the base material hardness).

Several investigations have been conducted to study the enhancement of superplasticity behaviour in friction stir processed alloys. In the FSP of Al-8.9Zn-2.6Mg-0.09Sc, Charit and Mishra et al. [15] reported a maximum superplasticity of 1165% at a strain rate of 3×10^{-2} s^{-1} and 310 °C with a grain size of 0.68 μm. More recently, F.C. Liu [16] reported a fine-grain micro-structure of 2.6 μm sized grains by applying FSP to extruded samples of an Al-Mg-Sc alloy, achieving a maximum elongation of 2150% at a high strain rate of 1×10^{-1} and a temperature of 450 ºC.

A ultrafine-grained FSP Al-Mg-Sc alloy was also reported [17] with a grain size of 0.7 μm exhibited high strain rate superplasticity, for a low temperature range of 200 to 300 °C with a single pass. For a strain rate of $3x10^{-2}$ s^{-1} at a temperature of 300 °C, a maximum ductility of 620% was achieved. However, for a temperature of 350 °C, abnormal grain growth was observed, as grain size increased and the samples no longer presented superplasticity, thus confirming that grain size is essential for the existence of a superplastic behaviour. García-Bernal et al. [18] conducted a study to evaluate the high strain rate superplasticity behaviour during the high-temperature deformation of a continuous cast Al-Mg alloy, having reported that the generation of a fine grain structure and the breaking of cast structure led to a significant improvement in its ductility up to 800% at 530 °C and a strain rate of 3x10-2 s-1.

The fine-grained microstructure generated by FSP can also prevent fatigue crack initiation and propagation due to the barrier effect of grain boundaries. For example, Jana et al. [19] friction stir processed a cast Al-7Si-0.6Mg alloy, widely used for its good castability, mechanical properties and corrosion resistance, but characterized by poor fatigue properties. The authors succeeded to improve fatigue resistance by a factor of 15 at a stress ratio of R=σ_{min}/ σ_{max}= 0 due to a significant enhancement of ductility and a homogeneous redistribution of refined Si particles.

Intense plastic deformation and material mixing featured in the FSP of A356 aluminium casting also resulted in the significant breakage of primary aluminium dendrites and coarse Si particles, creating a homogenous distribution of Si particles in the aluminium matrix and eliminating casting porosity [7]. This led to a significant improvement of ductility and fatigue strength in 80%, proving that FSP can be used as a tool to locally modify the microstructures in regions experimenting high fatigue loading.

Friction surfacing of AA6082-T6 over AA2024-T3 evidenced a significant improvement of wear performance in about 25 %, compared to the consumable rod in as-received condition (Table 1). This enhancement in wear behaviour is also due to a finer equixial grain microstructure within the coating, compared to the rod anisotropic microstructure which is more prone to delamination under wear loads.

AA2024-T3 substrate plates exhibited the best tribological properties, presenting the lowest weight loss, frictional force and friction coefficient. This is most likely due to both its higher surface hardness and its lower ductility, which make this material less prone to suffer plastic deformation under abrasive wear, in comparison with the AA6082 coating and the rod in as-received condition. Due to the fine grain structure observed, the coatings present high frictional force and coefficient (10.9 N and 0.56, respectively).

2.5. Surface composites

FSP has been also investigated to produce layers of hard materials on soft substrates, as aluminium based alloys. Most of the published work is focused on the effect of processing parameters on surface characteristics and techniques to evaluate the performance of modified surfaces. Nevertheless, the reinforcing particles deposition method is relevant in terms of

Material	Weight lost [mg]	Volume lost [mm³]	Volume rate [10⁻² mm³/m]	Wear rate [mg/m]	First stage Run-in wear		Second stage Steady state wear	
					Frictional force [N]	Friction coefficient	Frictional force [N]	Friction coefficient
Substrate	12.6 ± 3	4.54	1.51	0.042	4.9 ± 0.97	0.25 ± 0.05	7.5 ± 0.33	0.38 ± 0.017
ARCR	30.2 ± 5	11.19	3.73	0.101	-	-	7.1 ± 1.14	0.36 ± 0.059
Coatings	23.2 ± 3	8.59	2.86	0.077	-	-	10.9 ± 0.58	0.56 ± 0.029

Table 2. Weight loss due to wear (average values).

structural and chemical homogeneity and depth of the modified layer which influence the final surface performance. Different methods for depositing reinforced particles have been reported. A main reinforcing method consists of mixing reinforcing particles or powders with a volatile solvent such as methanol or a lacquer, in order to form a thin reinforcement layer, preventing reinforcing powders to escape. Another method consists of machining grooves in the substrate, pack these with reinforcing particles and process the zone with a non consumable FSP tool in a single pass or in multiple passes.

An enormous diversity of materials is used for surface reinforcements, the majority being hard ceramic particles as SiC, Al_2O_3 and AlN to improve surface properties as hardness, superplasticity, formability, corrosion and wear resistances.

The paricle size is relevant since small particles lead to higher concentration along bead surface and to smooth fraction gradients both in depth and along the direction parallel to the surface, while the thickness of the reinforced layer decreases with increasing particle size and is, typically, below 100 micron [20].

More recently, nanostructured layers have been produced and less common reinforcements were studied successfully. Two examples are: the incorporation of multi-walled carbon nanotubes (MWCNT) into a number of metallic materials as reinforcing fibres is a topic of recent interest due to the unique mechanical and physical properties of this material, namely very high tensile strengths [21]. FSP was tested to produce a composite of an aluminium alloy with MWCNT. Nanotubes were embedded in the stirred zone and the multi walled was retained. With tool rotational speeds of 1500 and 2500 rpm the distribution of nanotubes increased. Aiming at weight reduction of vehicles, FSP MWCNT/AZ31 surface composite were produced by Morisada et al. [22] and succeeded to disperse MWCNT into a AZ31 matrix. The microhardness increased to values of about 74 HV and the addition of MWCNT was seen to further promote grain refinement by FSP.

Another example is the incorporation of Nitinol (NiTi) that is a shape memory alloy with superelastic behavior and good biocompatibility. These alloys are widely used in orthodontics, but also in sensors and actuators. The possibility of incorporating wires, ribbons or powders into metallic matrixes opens up new applications for shape memory alloys. Studies report on the use of NiTi wires, but few have been made in the dispersion of NiTi powders in a metal matrix. Dixit et al. [23] produced a NiTi reinforced AA1100 composite using FSP and the particles were uniformly distributed. Good bonding with the matrix was achieved and no interfacial products were formed. The authors suggest that under adequate processing, the shape memory effect of NiTi particles can be used to induce residual stress in the parent matrix, of either compressive or tensile type. This study showed that samples had enhanced mechanical properties such as: Young modulus and micro hardness. A more recent work showed the possibility to introduce 1x2 mm ribbons of NiTi in AA1050 alloy by FSP showing a good vibration and damping capacity of the composite [24].

Shafei-Zarghani et al. [25] used multiple-pass FSP to produce a superficial layer of uniformly distributed nano-sized Al2O3 particles into an AA6082 substrate. Hardness was increased three times over that of the base material. Wear testing revealed a significant resistance improvement. Researchers also found that the increase of the number of passes leads to more uniform alumina particle distributions with a significant increase of surface hardness. The nano-size Al2O3 powder was inserted inside a groove with 4 mm depth and 1 mm width, which was closed by a tool with a shoulder and no pin.

3. Friction surfacing

3.1. Principles and process parameters

Friction surfacing (FS) was first patented in the 40's and is now well established as a solid state technology to produce metallic coatings. While FSP modifies the microstructure of a surface by simply deforming, recrystallize and homogenise the grain structure, FS modifies its chemistry. In friction surfacing a consumable rod under rotation is pressed under an axial load against the surface as depicted in Fig. 9. Heat generated in the initial friction contact promotes viscoplastic deformation at the tip of the rod. As the consumable travels along the substrate, the viscoplastic material at the vicinity of the rubbing interface flows into flash or is transferred over onto the substrate surface, while pressure and heat conditions triggers an inter diffusion process that soundly bonds the deposit. As the material undergoes a thermo-mechanical process, a fine grain microstructure is also produced by dynamic recrystallization.

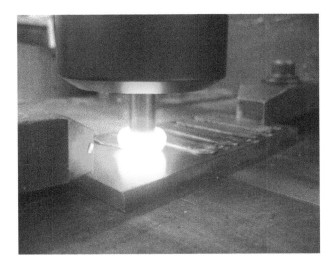

Figure 9. Metallic coating of steel substrate by FS

Gandra et al [20] proposed a model for the global thermal and mechanical processes involved during friction surfacing based on the metallurgical transformations observed when depositing mild steel over mild steel and is shown in Fig.10.

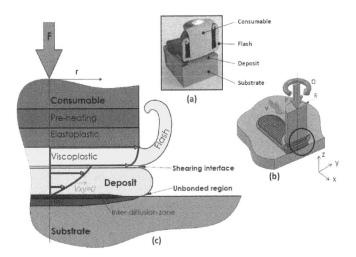

Figure 10. Thermo-mechanics of friction surfacing. (a) Sectioned consumable, (b) Process parameters and (c) Thermo-mechanical transformations and speed profile. Nomenclature: F – Forging force; Ω – rotation speed; v -travel speed; Vxy – rod tangential speed in-plan xy given by composition of rotation and travel movements [20]

The speed difference between the viscoplastic material, which is rotating along with the rod at v_{xy}, and the material effectively joined to the substrate ($v_{xy} = 0$), causes the deposit to detach from the consumable. This viscous shearing friction between the deposit and the consumable is the most significant heat source in the process.

Since the deposited material at the lower end is pressed without lateral confinement, it flows outside the consumable diameter, resulting into a revolving flash attached to the tip of the consumable rod and side unbounded regions adjacent to the deposit. Flash and unbonded regions play an important role as boundary conditions of temperature and pressure for the joining process.

Fig 11 shows typical material combinations tested using FS with successful results.

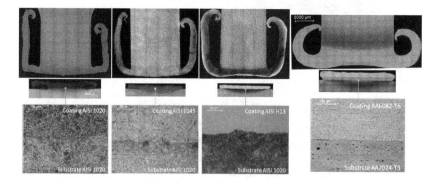

Figure 11. Different coatings/substrates combinations

The process allows the deposition of various dissimilar material combinations as the deposition of stainless steel, tool steel, copper or Inconel on mild steel substrates, as well as, stainless steel, mild steel and inconel consumables on aluminium substrates.

The influence of processing parameters on the deposit characteristics and bonding strength has been studied [26,27] aiming to correlate the resulting coating geometrical characteristics (thickness and bonded width) and mechanical performance with forging force, spindle and travel speeds. The increase of forging force improves the bond strength and reduces the coating thickness. The undercut region decreased when the forging force increased and the travel speed decreased. Higher ratios between the consumable rod feeding rate and the travel speed resulted in superior bonding quality. The applied load on the consumable rod was found to be essential to improve joining efficiency and to increase the deposition rate. Higher rotation or travel speeds were detrimental for the joining efficiency. Tilting the consumable rod along the travel direction proved to improve the joining efficiency up to 5 %. The material loss in flashes represented about 40 to 60 % of the total rod consumed, while unbonded regions were reduced to 8 % of the effective coating section in mild steel deposition. Friction surfacing was seen to require mechanical work between 2.5 and 5 kJ/g of deposited coating with deposition rates of 0.5 to 1.6 g/s, that is, deposition rates are higher than for laser cladding or plasma arc welding and the specific energy consumption lower than for other cladding processes.

In the friction surfacing of low carbon steel with tool steel H13 consumable rods, Rafi et al. [28] concluded that the coating width was strongly influenced by the rotation speed, while thickness was mostly determined by the travel speed.

This field of exploitation of producing aluminium coatings on aluminium based alloys is very promising. It was seen that friction surfacing enables intermediate mass deposition rates and higher energy efficiency in comparison with several mainstream laser and arc welding cladding processes. The required mechanical work varied between 2.5 and 5 kJ/g of deposited coating with deposition rates of 0.5 to 1.6 g/s. The forging force enhances joining quality while contributing to a higher overall coating efficiency. Faster travel and rotation speeds improved deposition rates and coating hardness, while decreasing energy consumption per unit of mass. Surface hardness increased up to 115 % compared to consumable rod. By adjusting a proper tilt angle, specific energy consumption drops, while slightly improving deposition rate and joining efficiency.

4. Summary and future trends

Friction based processes comprise Friction Stir Processing (FSP) and Friction Surfacing (FS).

Friction stir processing is mostly used to locally eliminate casting defects and refine microstructures in selected locations, for property improvements and component performance enhancement. Aluminium and steel castings are amongst the most common components improved by this technology aiming at eliminating porosities, destroy solidification structures with inhomogeneous segregated phases, refine grain structures improving n-service performance.

The recent advances in adding reinforcing particles to manufacture surface alloys and metal matrix composites is a breakthrough in this technology opening new possibilities to manufacture composites nanostructured with tremendous properties.

Friction surfacing has been used in the production of long-life industrial blades, wear resistant components, anti-corrosion coatings and in the rehabilitation of worn or damaged parts such as, turbine blade tips and agricultural machinery. Other applications feature the hardfacing of valve seats with stellite and tools such as punches and drills.

Since the deposits result from severe viscoplastic deformation, friction surfacing presents some advantages over other coating technologies based on fusion welding or heat-spraying processes, that produce coarse microstructures and lead to intermetallics formation, thereby deteriorating the mechanical strength of the coatings. However, friction surfacing currently struggles with several technical and productivity issues which contribute to a limited range of engineering applications.

Author details

R. M. Miranda[1*], J. Gandra[2] and P. Vilaça[2]

*Address all correspondence to: rmiranda@fct.unl.pt

1 Mechanical and Industrial Engineering Department, Sciences and Technology Faculty, Nova University of Lisbon, Caparica, Portugal

2 Mechanical Engineering Department, Lisbon Technical University, Av. Rovisco Pais, Lisboa, Portugal

References

[1] Bunshah RF. Handbook of Deposition Technologies for Films and Coatings: Science, Technology and Applications. 2nd Edition. Noyes Publications; 1994.

[2] Klopstock H, Neelands AR. Patent specification, An improved method of joining or welding metals; Ref. 572789; 1941.

[3] Nicholas ED. Friction surfacing, ASM handbook: Vol 6. ASM International; 1993; p. 321–323.

[4] Bedford GM, Vitanov VI, Voutchkov II. On the thermo-mechanical events during friction surfacing of high speed steels. Surface and Coatings Tecnhology 2001; 141 (1) 34-39.

[5] Thomas W. Friction Stir But Welding, International Patent Application Nº PCT/ GB92/02203 and GB Patent Application Nº 9125978.8, US Patent Nº5460317.

[6] Mishra RS, Ma ZY. Friction stir welding and processing. Materials Science & Engineering R 2005; 50 (1-2) 1-78.

[7] Nascimento F, Santos T, Vilaça P, Miranda RM, Quintino L. Microstructural modification and ductility enhancement of surfaces modified by FSP in aluminium alloys. Materials Science and Engineering A 2009; 506 (1-2) 16-22.

[8] Chang ADu, XH, Huang JC, Achieving ultrafine grain size in Mg-Al-Zn alloy by friction stir processing. Scripta Materiallia 2007; 57 (3) 209-212.

[9] Gandra J, Miranda R, Vilaca P, Velhinho A, Pamies-Teixeira J. Functionally graded materials produced by friction stir processing. Journal of Materials Processing Technology 2011; 211 (11) 1659-1668.

[10] Porter DA, Easterling KE, Phase Transformations in Metals and Alloys 1992, CRC Press.

[11] Gandra J, Miranda RM, Vilaça P. Effect of overlapping direction in multipass friction stir processing. Materials Science and Engineering A 2011; 528 (16-17) 5592–5599.

[12] Nakata K, Kim YG, Fuji H, Tusumura T, Komazaki T. Improvement of mechanical properties of aluminium die casting alloy by multi-pass friction strir processing. Materials Science and Engineering A 2006; 437 (2) 274-280.

[13] Santella ML, Engstrom T, Storjohann D, Pan TY. Effects of friction stir processing on mechanical properties of the cast aluminum alloys A319 and A356. Scripta Materialia 2005; 53 (2) 201-206;

[14] Feng AH, Ma ZY. Enhanced mechanical properties of Mg-Al-Zn cast alloy via friction stir processing. Scripta Materiallia 2007; 56 (5) 397-400.

[15] Charit I, Mishra RS, Low temperature superplasticity in a friction-stir-processed ultrafine grained Al-Zn-Mg-Sc alloy. Acta Materialia 2005; 53 (15) 4211-4223.

[16] Liu FC, Ma ZY. Achieving exceptionally high superplasticity at high strain rates in a micrograined Al-Mg-Sc alloy produced by friction stir processing. Scripta Materialia 2008; 59 (15) 882-885.

[17] Liu FC, Ma ZY, Chen LQ. Low-temperature superplasticity of Al-Mg-Sc alloy produced by friction stir processing. Scripta Materialia 2009; 60 (5) 968-971.

[18] García-Bernal MA, Mishra RS, Verma R, Hernández-Silva D, High strain rate superplasticity in continuous cast Al-Mg alloys prepared via friction stir processing. Acta Materialia 2009; 60 (10) 850-853.

[19] Jana S, Mishra RS, Baumann JB, Grant G. Effect of stress ratio on the fatigue behavior of a friction stir processed cast Al-Si-Mg alloy. Scripta Materialia 2009; 61 (10) 992-995.

[20] Gandra J, Miranda RM, Vilaça P. Performance Analysis of Friction Surfacing. Journal of Materials Processing Technology 2012; 212 (8) 1676-1686.

[21] Arora HS, Singh H, Dhindaw BK. Composite fabrication using friction stir processing - a review. International Journal of Advanced Manufacturing Technologies 2012; 61 (9-12) 1043–1055.

[22] Morisada Y, Fujii H, Nagaoka T, Fukusumim M, MWCNTs/AZ31 surface composites fabricated by friction stir processing. Materials Science and Engineering A 2006; 419 (1-2) 344–348.

[23] Dixit M, Newkirk JW, Mishra RS. Properties of friction stir-processed Al1100-NiTi composite. Scripta Materialia 2007; 56 (6) 541-544.

[24] Mendes L. Production of aluminium based metal matrix composites reinforced with embedded NiTi by friction stir welding. MSc thesis, Universidade Nova de Lisboa, 2012

[25] Shafei-Zarghani A, Kashani-Bozorg SF, Zarei-Hanzaki. Microstructures and mechanical properties of Al/Al2O3 surface nano-composite layer by friction stir processing. Materials Science and Engineering A 2009; 500 (1-2) 84-91.

[26] Vitanov VI, Voutchkov II, Bedford GM. Decision support system to optimize the Frictec (friction surfacing) process. Journal of Materials Processing Technologies 2000; 107 (1-3) 236-242.

[27] Vitanov VI, Javaid N, Stephenson DJ. Application of response surface methodology for the optimisation of micro friction surfacing process. Surfaces and Coatings Technology 2010; 204 (21-22) 3501-3508.

[28] Khalid Rafi H, Janaki Ram GD, Phanikumar G, Prasad Rao K. Microstructural evolution during friction surfacing of tool steel H13. Materials and Design 2011; 32 (1) 82-87.

Chapter 3

Modern Orthopaedic Implant Coatings — Their Pro's, Con's and Evaluation Methods

Jim C.E. Odekerken, Tim J.M. Welting,
Jacobus J.C. Arts, Geert H.I.M. Walenkamp and
Pieter J. Emans

Additional information is available at the end of the chapter

1. Introduction

Metal implants in the field of orthopaedic and trauma surgery have been used for a long time to restore joint function, reduce pain or stabilize fractures [1-3]. Both prevention of infection as well as integration of the implant with the host-tissue (mostly bone) are factors of concern. The most frequently used implants for these purposes are made of metallic alloys (for plates and nails for fracture repair and for total joint arthroplasties) and polyethylene (PE) (used in the articulating parts of a prosthesis).

Infection of orthopaedic implants and prostheses is a medical issue that has intrigued people since their very first use over two-and-a-half millennia ago. Since that early beginning, lack of fixation and infection has been the major problem with such medical devices. In many cases, it may result in serious discomfort, limb amputation, illness and in many cases it may have even resulted in death of the patient. Even after 2500 years of medical progression we are still not able to fully conquer these major health risks. Since they are not isolated to the field of orthopaedics and trauma, their multi-factorial character indicates the complexity of the problems concerning healthcare-associated infections (HAI) [4, 5].

The purpose of this chapter is to give an insight in the quest to decrease the percentage of infections and increase the amount of osteointegration by coatings on metallic alloys in the field of orthopaedic and trauma surgery. Finally an overview will be provided of the available methods to examine and evaluate the properties of coatings *in vitro* and *in vivo*.

1.1. Healthcare-associated infections and orthopaedics

Healthcare-associated infections, also called nosocomial infections, are considered to be the biggest healthcare related complication worldwide. HAI annually affects over 600 million patients worldwide with approximately 4.1 million patients in Europe and about 1.7 million patients in the United States [6, 7]. These infections can be related to the cause of death of a considerable number of patients annually. Together with the tremendous economic burden of HAI, HAI is a major point of interest in medical research.

Area	# Patients (million/yr)	Prevalence (%)	Death (%)	Costs (billion)	Neonatal death rate (caused by HAI)
Europe	4.1	7.1	3.7%	(€) 7	< 5%
USA	1.7	4.5	5.8%	($) 6.5	< 5%
Worldwide	> 600	8.5 - 15.5			Up to 75% in South-East Asia and Africa

Table 1. Epidemiological data on HAI [6, 7].

With urinary tract infections as the most frequent implant related HAI in developed countries, orthopaedic implant infections is another major sub-populations within the multifactorial group of HAI (together with infections related to cardiovascular, neurological and gastrointestinal interventions). Infections due to implantation of total hip and total knee prostheses account for about 2% of the HAI, without taking trauma implants into account [7]. Trauma implants or implants for fracture fixation and stabilization, like plates, screws and stabilizing frames, have been described to have an even higher risk for infection, mainly due to the fact that they are used to repair complex injuries and open fractures. Infection together with the eventual loosening of an orthopaedic implant explains the limited lifespan of an orthopaedic device (generally up to 15 years for an artificial joint [5, 8, 9]).

Since the discovery of antibiotics, (implant) infections have been reduced and implant infections have become less lethal and can even be cured. Still, the extensive use of antibiotics has resulted in an increasing amount of resistant bacterial strains, which makes infections caused by those pathogens challenging. Medical device implantation remains troublesome also in the case of orthopaedic implants.

1.2. The race for the surface

After implantation of an orthopaedic device, a competition between bacterial colonization versus tissue integration takes place to conquer the surface of the implant. This phenomenon is often described as "the race for the surface" (Figure 1) [10, 11].

The race for the surface

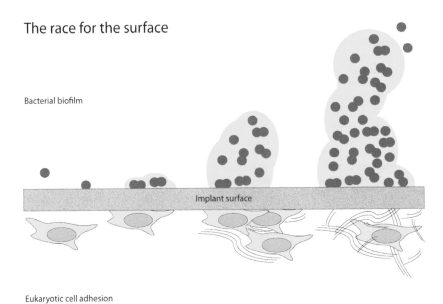

Bacterial biofilm

Implant surface

Eukaryotic cell adhesion

Figure 1. A schematic representation of "the race for the surface", between the bacterial biofilm colonization and eukaryotic cell adhesion with subsequent bone apposition on the implant surface.

The first stage of bacterial biofilm formation is the settling of a planktonic bacterium on the surface of the implant. After adhering to the surface, the bacterium starts to divide and encapsulate itself for protection against the host organism's immune response. This layer of protective matrix, mostly consisting out of polysaccharides, also shields the bacteria from effective antibiotic treatment. The first stage of the biofilm formation is complete and subsequently the present bacteria start to form colonies increasing the internal pressure in the biofilm, which starts to expand. At a certain point the bacterial load within the mature biofilm becomes so high that planktonic bacteria are released from the biofilm. These bacteria can then result in the infection of the surrounding tissue or in the expansion of the biofilm on a different location (Figure 1) [10-12]. Eukaryotic cell adhesion (e.g. adhesion of osteoblasts) on the other hand, can result in implant ingrowth by settling of the osteoblast on the implant surface, followed by cell division and collagen matrix production. Eventual calcification of the collagen matrix allows bone apposition on the implant surface (Figure 1) [10, 12]. In general the inability of the body and its immune system to cope with infected implants is one of the biggest issues when implants are used for medical treatment. Due to infection, local bone resorption takes place, leading to bone loss and implant loosening. As such it is essential to treat the infection, avoiding the risk of tremendous damage to the bone and the bony peri-implant tissue. After removal of an infected implant, the accompanying bone fractures, soft tissue infection, and inflammation result in fixation issues and an increased infection risk during revision surgery [13].

1.3. Implant coatings

In order to decrease the amount of implant infections and prevent the implants from loosening, coatings can be applied to the surface of the implant. These coatings may vary from (antibiotic) releasing to non-releasing coatings. In general non-releasing coatings (like hydroxyapatite) are applied by thermal-processes, while releasing coatings (like RGD or antibiotic-containing coatings) are mostly applied to the surface by dip or spray coating, due to their limited thermal stability.

Since the principle of the "race for the surface" dictates that early tissue integration may also reduce the infection risk, a coating promoting tissue integration may also be regarded as a passive method to reduce the amount of infections. In order to promote this tissue integration, one of the biggest leaps forward in the improvement of implant fixation and "the race for the surface" in favour of eukaryotic cells might be the use of hydroxyapatite (HA) coatings on the surface of a metallic implant [12, 14-16]. Although in the beginning it was believed that uncemented prostheses, including the HA-coated implants, would have a higher infection percentage compared to implants fixated with an antibiotic-releasing bone-cement, long-term studies showed a comparable infection percentage and a longer survival in favour of the uncemented prosthesis [5, 17]. These HA-based coatings (and their derivatives) are still one of the most frequently used implant coatings in the field of orthopaedic surgery and trauma, resulting in improved implant ingrowth and a longer lifespan of the prosthesis [8]. A combined situation of a coating with both antimicrobial and osteoconductive properties, is yet to be found.

2. Active and passive implant coatings

2.1. Osteoconductive coatings

The definition of osteoconduction means that a material or coating "guides" the bone healing or formation. In case of coatings this means that the bone formation is "guided" to grow or attach to the coating surface (a passive coating) [18]. An orthopaedic implant with such a surface treatment or coating provides an ideal substrate for bone aposition which results in improved implant fixation and a possible prolonged lifespan, with a decreased risk of implant loosening and possibly infection [10, 12].

2.1.1. Apatite coatings

The initial idea of hydroxyapatite (HA) coatings originated from the use of calciumphosphates as a material to stimulate osteogenesis, like tricalciumphosphate (TCP). During the last decades, studies have shown that HA and TCP are suitable for the production of ceramic scaffolds to serve as a bone substitute. Studies on the stability of sintered TCP ($Ca_3(PO_4)_2$) and

HA ($Ca_{10}(PO_4)_6(OH)_2$) have shown that TCP ceramics dissolve over 10 times faster in acidic and alkaline environments compared to HA [19]. Explaining the rationale behind the current use of TCP for resorbable bone scaffolds and the use of HA for implant coatings.

Since the proposed use of HA as a coating, in the late 1980's by Geesink*et al.*[14-16], several implant designs have been used in the clinic, e.g. partially coated or fully coated hip implants. Fully coated implants achieved complete bone remodeling around the implant, with very good fixation properties. The major disadvantage of these fully coated implants was that in case of revision surgery (either for implant infection or component failure) removal of the implant resulted in massive bone trauma due to the fixation of the implant to the bone. By redefining the coating location to the taper only, a good fixation could be achieved with limited problems at the time of a revision surgery [15]. However this design allowed the formation of stress shielding due to the pressure of the stem against the cortical wall. Due to this strain an implant can get loose, resulting in bone loss or a cortical wall fracture. Still HA coatings sintered to an implant surface has proven itself to be the most successful implant coating made, with 20 years of clinical experience [8, 20].

2.1.2. Hydroxyapatite application methods for metallic surfaces

There are several ways to coat a metallic alloy, like titanium or stainless steel, with HA. The techniques to coat such a metallic implant include; dip coating [21], sputter coating [22], pulsed laser deposition [23], hot pressing and hot isostatic pressing [24], electrophoretic deposition [25], electrostatic spraying [26, 27], thermal spraying [28], and sol–gel [29]. Some of these techniques are still experimental, thermal spraying, in particular. Plasma spraying is the most accepted method for the production of HA coatings [30]. Plasma spraying requires high temperatures which may damage the HA crystallinity and create unwanted or amorphous phases, with HA ablation from the coated surface as a possible result [28]. Every technique has its advantages and disadvantages. For example, the thickness, the bonding strength and the properties of the HA-composition may be influenced by the application technique. Techniques such as thermal spraying and sputter coating are used for surfaces or substrates (e.g. porous titanium implants) which are difficult to coat. Techniques such as electrophoretic deposition and sol-gel may coat more complex substrates such as porous alloys, still the production of crack free coatings remains challenging (Table 2).

2.2. Osteoinductive coatings

Although biomimetic HA coatings improve the osteoconductivity of metal implants, they do not influence the osteoinductivity. In general osteoinductive coatings are described as coatings which induce bone formation of undifferentiated cells in the surrounding tissue to ultimately promote osteointegration of bone to the coating (active coatings). In order to promote the differentiation of immature progenitor cells to an osteoblastic lineage, attempts to integrate functional biological agents such as growthfactors into biomimetic coatings have been realized [33, 34]. Several of these coatings have been studied extensively, the most important coatings are described below.

Technique	Advantage	Disadvantage	Thickness	Ref
Sol-Gel	• Coats 3D complex porous substrates • Low processing temperatures • Relatively cheap • Very thin coatings	• Processing in controlled atmosphere	< 1 µm	[29]
Sputter coating	• Uniform coating thickness on flat substrates	• Only coats visible area • Expensive and time consuming • Unable to coat complex 3D porous substrates • Risk for amorphous coating	0.02 - 2 µm	[22]
Pulsed Laser Deposition	• See sputter coating	• See sputter coating	1 - 10 µm	[23, 25]
Electrostatic Spray Deposition	• Uniform coating thickness on flat substrates • Relatively cheap	• Only coats visible area • Fragility	1 - 10 µm	[26, 27, 31]
Electrophoretic Deposition	Uniform coating thickness • High deposition rates • Coatscomplex 3D poroussubstrates	• Cracks in coating • High sintering temperatures	0.1 - 200 µm	[25]
Thermal spraying, Plasma spraying	• High deposition rates	• Only coats visible area Coating decomposition due to high temperature • Rapid cooling may result in amorphous coating	30 – 200 µm	[25, 28]
Dip Coating	• Inexpensive Coatings applied quickly • Coatscomplex 3D poroussubstrates	• High sintering temperatures • Thermal expansion results in amorphous coating • Fragile due to thickness	0.05 - 2 mm	[21, 25]
Hot Pressing, Hot Isostatic Pressing and Sintering	• Dense coatings	• Unable to coat complex 3D porous substrates • Differences in coating elastisity • Expensive • Interaction with or changes due to the encapsulation material	0.1 – 10 mm	[24, 25, 32]

Table 2. Different coating techniques to apply hydroxyapatite (HA) on an implant.

2.2.1. RGD coatings

Extracellular matrix proteins contain a short functional domain of three aminoacids, arginine (R), glycine (G) and asparagine (D), the so-called RGD-domain. This domain plays an important role in cell adhesion, cell proliferation and differentiation. RGD peptides coated to a surface can initiate these processes in their direct vicinity. The major advantage of using peptide coatings instead of protein coatings is that peptides are smaller and more stable compared to proteins. This allows more peptides to be coated to a surface, which results in a more dense coating. Studies have shown that the flanking amino acids in a RGD containing peptide are of great importance for their efficiency. *In vitro* studies show promising results, where RGD enhances (human) cell adhesion, proliferation and differentiation in the osteogenic lineage [35]. An *in vivo* study of an HA/HA-RGD coating with antibiotic release showed that the HA-RGD coating performed at least equally well as the HA-only coating [33]. Still these coatings remain experimental by application.

2.2.2. BMP coatings

Bone morphogenic proteins (BMP's), which belong to the transforming growth factor-β (TGF-β) superfamily, are generally accepted osteoinductive additives for per-operative use to enhance bone remodeling. Due to the lack of a local delivery system, capable of a sustained release, relatively high dosages of BMP's (e.g. BMP-2) are being used in the clinic. The use of BMP's has a locally higher incidence of tumorigenesis as a major disadvantage. Other osteogenic BMP's, such as BMP-4 [36, 37] and BMP-7 [38], are also potent inducers of bone regeneration. Knippenberg et al showed that BMP7, in contrast to BMP2 may be a more chondrogenic growth factor in contrast to BMP2 which was described as a more osteogenic growthfactor [39]. BMP-2 works in 2 concentration dependent directions, at low dosages it promotes bone remodeling, while at high dosage it promotes bone resorption. Therefore a low-dose releasing BMP-2 coated implant may be the most optimal [40]. While many techniques to incorporate BMP's result in a burst release, they can also be incorporated into the crystal latticework of coatings to establish a gradual release system [34, 40]. As such the incorporated proteins can be released gradually and steadily at a low pharmacological level; not rapidly as in a single high-dose burst [41]. In conclusion, incorporation of osteoinductive coatings may seem attractive but, release rate, potential carcinogenesis, inactivation of the compound (e.g. due to changes in temperature, pH), and bonding to the implant remain of concern.

3. Antimicrobial coatings

Since vascularization of infected tissue is often compromised and a bacterial biofilm is formed, which results in poor penetration of antibiotics, systemic antibiotics are not sufficient to treat a bone infection properly [10-12]. To achieve a local dose high enough to treat the infection, this would involve a systemic overdose of the antibiotic (possibly resulting in kidney and liver failure and damage to the function of the inner ear). The best solution to this problem is to have a local delivery system, this suggests an approach for the treatment of orthopaedic infections. Still in many cases the prosthesis can be rescued by infection treatment *in situ*, without a surgical procedure [13, 42, 43]. The use of local antibiotics by antibiotic-loaded bone cement, either as beads or spacers often placed after implant removal in the remaining infected cavity. The general treatment procedure requires at least two surgical procedures, one to remove the infected implant and the surrounding affected tissue, combined with the placement of a spacer or antibiotic-loaded beads to fill up to void [44-46]. The second operation is required to remove the spacer or beads after a couple of weeks or months. Once the infection is regarded as treated sufficiently, a new implant or prosthesis is implanted. If the treatment was not successful, new beads can be placed, which will require a third operation for the removal of the beads [44-46]. Due to the high costs and the tremendous burden for the patient a one-step procedure would be preferable. An antimicrobial coating directly on the surface of the newly placed implant, in case of revision surgery after infection, could prevent the infection form re-occurring, but such a coating may also work as a prophylactic in the case of the placement of a primary hip.

3.1. Antibiotic releasing coatings

Already in clinical use in other medical specialties (e.g. in sutures and central venous and urinary tract cathethers), antibiotic releasing coatings remain mainly experimental in the field of orthopaedic and trauma surgery. For orthopaedic applications gentamicin, vancomycin, rifampicin, and tobramycin are the most frequently used local antiobiotics in case of an implant infection. There are several published *in vitro* and *in vivo* studies based on the use of these antibiotic drugs for an orthopaedic implant coating. Poly-L-lactide (PLLA) coatings with rifampicin on a fracture fixation plate, placed on the tibia of rabbits, showed good results on both antimicrobial properties and acceptance of the host-tissue within 28 days after surgery [47]. Also the direct application of minocycline and rifampicin on titanium, placed in the distal femur of a rabbit, lead to good results on prevention of device colonization and infection prevention within a week after surgery [48]. A combined osteoconductive/antimicrobial coating (HA/tobramycin) on titanium, evaluated in the proximal tibia of a rabbit indicated the potential of a combined coating for infection prevention as well as implant incorporation [49]. A recent study on a combined osteoconductive/osteoinductive/antimicrobial coating (HA/RGD/gentamicin) on stainless steel showed promising results on bone integration and antibiotic release characteristics [33]. Furthermore antibiotic releasing coatings on biodegrad-able substances could replace antibiotic containing PMMA-beads, in this case no implant coating would be necessary. A study on gentamicin coated poly(trimethylene carbonate) (PMTC), a biodegradable polymer, showed good results on antibiotic release, biofilm inhibi-tion and biodegradability, suggesting to be a good substitute for PMMA-beads [50]. A recent report on a prospective study of the first antibiotic releasing tibial nail has shown promising clinical results with no deep surgical wound infections within the first six months after implantation [51]. The major disadvantage for these coatings which they will face in the near future is the increasing number of antibiotic-resistant bacterial strains. This is the main reason why antimicrobial coatings, based on disinfectants or non-traditional antibiotics, are of great interest in the research and development of such coatings.

3.2. Silver-based coatings

Silver is (amongst copper, lead and mercury) a potent antimicrobial heavy metal which has been related to medicine for many centuries. Instead of its metallic state, only the ionic state of silver is considered to be antimicrobial and its mode of function is multifactorial. Ionic silver not only reacts easily with amines and microbial DNA to prevent bacterial replication, but also with sulfhydryl groups of metabolic enzymes of the bacterial electron transport chain, resulting in their inactivation [52, 53]. This also forms its treat to large scale clinical applications, since it can also inhibit eukaryotic metabolic function in a patient. Therefore a local release of silver ions is preferable. In contrast to lead and mercury silver does not appear to have cumulative toxic effects in the body, suggesting its potential as a coating component. The use of silver in releasing coatings currently spans from central venous catheters to urinary tract catheters and coated orthopaedic implants, with limited *in vivo* antimicrobial effectiveness as a main problem. While some studies show that a silver coated surface can minimize the infection risks by lowering the bacterial load [54-57], to date, pre-clinical studies and randomized controlled trials of silver coated catheters, implants and external fixation pins were not able to prove its antimicrobial efficacy [52, 58-61].

4. Coating evaluation

Newly developed coatings need evaluation before implementation in the clinic to prevent possible adverse reactions to the coating. This evaluation includes mechanical testing and cytotoxicity and biocompatibility tests. In general these tests can be subdivided in two categories: *in vitro* and *in vivo* testing.

4.1. *In vitro* evaluation

This is defined as all testing modalities performed in controlled laboratory conditions, so outside of a living organism or its natural setting (Table 3).

4.1.1. Cytotoxicity tests

Cytotoxicity tests can be subdivided in cell viability, cell adhesion and cell spreading assays and are usually performed with fibroblastic cell lines (e.g. A529 [62], MC3T3-E1 [62-65], L929 [66], MG-63 [67, 68]). Cell viability assays evaluate the toxicity of a compound present in the vicinity of the cells either in solution or in a solid state. During these tests the material to be tested is incubated in cell culture medium. The resulting pre-conditioned culture medium is then used for cell-culture to evaluate the viability of the cells after exposure to the extracted medium from the material to be tested. Depending on the material, also direct cell culture on the material surface can be performed. The viability of the cells can e.g. be assessed by performing an MTT-assay.

• **The MTT-assay** is based on the reduction of 3-(4,5-dimethylthiazol-2-yl)-2,5-diphenyltetrazolium bromide (MTT, or another tetrazolium salt) to formazan by the enzyme succinate dehydrogenase in the mitochondria of living cells. The formed purple product can be measured spectrophotometrically and provides a direct measurement of the cell viability based on mitochondrial activity, hence energy metabolism [55, 64-70].

Analytical method	Detection method	Detects	Ref.
Eukaryotic cultures			
Tetrazolium based assay • MTT • XTT • MTS	Spectrophotometric	Cell viability by metabolic activity • Reduction of a tetrazolium salt (yellow) to formazan (purple) by metabolically active cells	[55, 64-70]
FDA based assay • Fluorescein diacetate • DAPI	Fluorescence	Cell adhesion • Fluorescein diacetate→ cytoplasm of healthy cells (green) • DAPI → nuclei of every cell (blue)	[64, 70, 71]
Crystal violet	Spectrophotometric	Cell viability by DNA content • DNA staining, released dye indicates level of cell viability compared to control situation	[62]
SRB • Sulforhodamine B	Spectrophotometric	Cell density based on protein content • protein staining, released dye indicates amount of cells present compared to control situation	[72]
Phalloidinbased assay • Rhodamine • DAPI	Fluorescence	Actin staining • Rhodamine → actin cytoskeleton (red) • DAPI → nuclei (blue)	[64, 68, 69, 73]
Alizarin Red S	Fluorescence	Osteogenic cells → staining of calcium deposition (red)	[62]
ALP • Alkaline phosphatase	Spectrophotometric • Enzymatic assay	ALP activity is a marker for osteogenic potential of a cell • Enzymatic turnover of p-nitrophenyl phosphate to p-nitrophenol	[62, 63, 69, 74]
Live/dead staining • Fluorescein diacetate • Ethidium bromide	Fluorescence	Cell viability • Fluorescein diacetate → cytoplasm of healthy cells (green) • Ethidium bromide → nuclei of death cells (red)	[65, 70]

Table 3. *In vitro* analytical methods – part 1

Analytical method	Detection method	Detects	Ref.
Prokaryotic cultures			
JIS Z 2801	Bacterial growth	Bacterial growth inhibition	[70, 73]
ASTM E-2810	Bacterial growth	Bacterial growth inhibition	[70]
Agar diffusion	Bacterial growth • Colony formation	Zone of inhibition, antibiotic potential of test compound • Distance antibiotic containing object to colony defines potency of antibiotic compound and its release system	[65]
MIC-MBC-assay	Bacterial growth • OD 600 measurements (MIC) • Quantitative bacterial culture (MBC)	Bacterial growth inhibition • Elevation in OD 600 indicates bacterial growth • Colony formation	[54, 55, 65, 69]
Other			
Immunocytochemistry	Light microscopy	Tissue specific staining	[62, 75]
Optical imaging	Fluorescence or bioluminescence • Fluorescence → emission after excitation • Bioluminescence → auto-emission	Presence of light emitting cells (e.g. cell growth or biofilm) • Fluorescence → GFP • Bioluminescence → luciferase	[76]
PCR	Fluorescence • SYBR-green related dyes	DNA/RNA expression	[62, 63]
Western blot	Chemiluminescence	Protein expression	[63]
SEM/TEM • Scanning electron microscopy • Transmission electron microscopy	Electron microscopy • Sputtering with gold or carbon for visualization	Evaluation of bacterial biofilm or extracellular matrix formation • SEM → Cell to surface interactions • TEM → Cell to cell interactions	[62-64, 66, 67, 73, 74]

Table 3. *In vitro* analytical methods – part 2

- **Cell adhesion assays** evaluate the potential of an implant surface to allow cell adhesion by culturing cells directly on the surface of the material to be tested. After allowing the cells to adhere to the surface, non-adhering cells are washed of the implant surface after which the adhering cells are double-stained with fluorescein diacetate (FDA) and ethidium bromide. In this live/death staining, FDA will stain the cytoplasm of intact cells green, while ethidium bromide will stain the DNA of dead cells red. The cell adhesion can be assessed by fluorescence microscopy. The ratio between the FDA-positive and ethidium bromide-positive cells provides insight into the live-dead percentage and thus biocompatibility of the culture surface. If the cells are only incubated with FDA, cell lysis allows quantification by fluorescence spectrophotometry. The level of fluorescent signal is an indication of cell adhesion on the material surface [64, 70, 71].

- **Cell spreading assays** evaluate the potential of a surface to allow cell adhesion and proliferation including matrix formation in the case of e.g. osteocytes. There are multiple ways to assess this surface property. One of the methods described is the use of cell staining directly on the surface after cell culture on the material surface by e.g. crystal violet staining or by an actin staining based on phalloidin. The crystal violet staining is a DNA staining in which cells are fixed on the cultured surface, then incubated with crystal violet to stain the cellular DNA. After washing the stained cells the dye is released by the incubation in a weak acid. The released dye can be measured on a spectrophotometer and providing a quantitative measure for the amount of cells present on the surface. A phalloidin-based staining allows staining of the actin cytoskeleton and cellular organization on the surface of a material. This is a direct cell staining which is visualized by fluorescence microscopy. In most cases the phalloidin based stainings are counterstained with DAPI to stain the cells nuclei, which allows visualization of the individual cells and their cytoskeleton [62, 64, 68, 69, 73].

- **Assays to assess the osteogenic potential,** quantify the osteogenic potential of a coating. This can be determined by using cultured cells on the coating surface for an alkaline phosphatase assay (ALP). The ALP assay determines the ALP activity within the tissue, which is related to osteogenesis and bone deposition on the coating surface. Another method to assess the osteogenic potential of a coated surface is an alizarin red s staining, which stains calcified tissue [62, 63, 69, 74].

4.1.2. Antimicrobial coating tests

In the case of antimicrobial coatings the effect of the coating on bacteria can be assessed with a wide variety of assays, with the most well-known being the agar diffusion test where the release of an antimicrobial compound into the agar leads to an inhibition zone around the releasing material.

- **Bacterial viability** can be assessed by a minimal inhibitory concentration (MIC)/ minimal bactericidal concentration (MBC) assay. In this assay the releasing material is allowed to release its effective compound into a buffer or culture medium over a certain time span. The acquired pre-conditioned buffer/medium is then used in a bacterial culture setting in which a standardized amount of bacteria is exposed to the preconditioned buffer/medium. After 24 hours of incubation the optical density can be measured at 600 nm, the lowest concentration which shows no increased optical density compared to the uncultured control condition defines the MIC, while the lowest concentration which shows no bacterial growth after incubation of the MIC-cultures on agar plates for another 24 hours defines the MBC [55].

- **In vitro biofilm formation** on a surface can be confirmed by incubating the surface in a bacterial suspension, rinsing the surface with an isotonic buffer (PBS) and use sonication to release the bacteria from the surface for quantitative culture. Or fix the bacteria on the surface with 2.5% glutaraldehyde/PBS for evaluation with SEM. This method can easily be transferred to the *in vivo* situation.

- **International standards** provide guidelines of how to assess coating stability and function, e.g. ISO 10993-5 provides guidelines for *in vitro* medical device evaluation. The Japanese Industrial Standard Z 2801 (JIS) describes a test for contact killing by the incubation of bacteria on a potential antimicrobial surface. Culturing of this surface provides insight on the antimicrobial properties of the evaluated coating [70, 73]. The American Standard E-2810 (American Society for Testing and Materials, ASTM)) describes a test for contact killing by the application of a bacteria loaded agar onto the coated surface, after incubation the number of viable bacteria is determined [70].

4.2. *In vivo* evaluation

This is defined as all testing modalities performed in a controlled group of living organisms, often including clinically relevant parameters and a broad range of imaging techniques (Table 4).

The first models concerning orthopaedic conditions date back to the late 19th century, primarily focusing on osteomyelitis [77]. Rodet described 2 basic methods to establish an osteomyelitis in a rabbit, the first one by inflicting a fracture and subsequent intravenous injection of the bacterium, resulting in an osteomyelitic leasion in the area of the fracture. The second method was performed by merely injecting bacteria intravenously, which resulted in a systemic infection with periosteal leasions leading to local osteomyelitis [77].

The most well-known model for osteomyelitis is the model by Norden *et al.*; this model describes the direct percutaneous injection, directly into the tibial intramedullary cavity of a rabbit, of both a scleroting agent (sodium morrhuate, bile salts from codfish) and *S. aureus* [78]. Andriole *et al.* however, established one of the first osteomyelitis models with a foreign object. Their model was based on a tibial fracture and subsequent tibial stabilization by a stainless steel pin, contaminated with *S. aureus* [79]. Together with the model by Norden, the model of Andriole mainly form the basis for current animal models used for the evaluation of implant coatings. In general, rabbits are still the most frequently used animal species for these experimental studies, but there have also been successful models in mice, rats, dogs and sheep. During the years, models increased in complexity and included multifactorial parameters.

The bone bonding properties of apatite-coated implants were first evaluated in dogs by Geesink *et al.* [16]. After the development of these apatite-coated implants, Vogely *et al.* described a rabbit proximal tibial model for the evaluation of hydroxyapatite coated titanium implants in an implant site infection [80]. Darouiche et al. were one of the first who described a rabbit lateral femoral condyle model for the evalauation of antimicrobial coatings on titanium [48]. Poultsides *et al.* described a haematogeneous implant contamination model by MRSA [81]. Moojen et al. evaluated a combined coating with both osteoconductive (periapatite) and antimicrobial (tobramycin) properties in a proximal tibial implant infection model in rabbits [49]. Also, Moskowitz *et al.* developed antibiotic multilayer implant coatings with an antibiotic release of over 4 weeks in a 2 stage rabbit distal femoral condyle infection model. The first surgical stage contained the initial infection with the insertion of a pre-colonized peg, the second surgical stage was the removal of the peg and implantation of the antibiotic coated implant [65]. Alt *et al.* was one of the first to describe a coating which combined osteoconductive (hydroxyapatite), osteoinductive (RGD) and antibiotic (gentamicin) properties in an experimental rabbit implant infection model [33].

4.2.1. Clinical parameters

Body weight and temperature provide general information about the animal's physical condition, with weight loss and fever in case of an infection. Leucocyte differentiation provides a detailed overview of the percentages of lymphocytes, neutrophillic granulocytes, monocytes, basophilic granulocytes and eosinophilic granulocytes in the total leucocyte population. An elevated number of leucocytes or a shift in differentiation indicates a bacterial infection. The ESR is based on the fibrinogen balance in the blood. In case of an inflammation or infection the fibrinogen levels increase, resulting in agglutination of erythrocytes with sedimentation as a result. CRP is an acute phase protein whose levels rapidly increase in case of inflammation or infection. ESR and CRP both only indicate the presence of inflammation or infection, never the cause or the location [82].

4.2.2. Imaging modalities

- **Optical imaging (based on fluorescence and bioluminescence)** is based on the detection of light emitted from the body. The use of fluorescently labeled antibodies results in a very specific signal, although resulting in a very local detection, it also requires a large amount of antibodies when used in humans. This renders large scale use in humans not yet profitable [76]. Also bioluminescence can be used to visualize infection. The main disadvantage of bioluminescence is the requirement of the luciferase gene in the cell to be detected, meaning the use of genetically modified organisms in case of detection by either autologous cells or bacteria. E.g. a bacterium expressing luciferase can be used to monitor an implant infection initiated with that bacterium [76]. Both methods are currently available in laboratory animal setting.

- **X-ray** is by far the oldest imaging technique and the most frequently used imaging technique to assess fractures, implant fixation, location and loosening, but also for the differential diagnosis of bone diseases like osteomyelitis. The use of X-rays is cost effective, they are easy to obtain and have a relatively low burden for the patient. An X-ray only provides detailed information about the mineralized tissue (or the lack thereof) and the disease related changes accompanied with it [15, 83, 84].

- **CT (computed tomography)** is a 3D-imaging technique which uses X-rays to construct a 3D image of the mineralized tissue in a patient. It generally provides more in-depth data about the density of the mineralized tissue and bone remodeling compared to X-rays, however imaging of metallic implants can result in scattering of the X-rays resulting in a blur around the implant, rendering data-analysis difficult [76, 85, 86].

- **DEXA (dual energy X-ray absorptiometry)** is often incorrectly stated as a bone scan. The use of 2 different energy levels of the X-ray beam allows accurate determination of the bone mineral density. DEXA is the most common imaging technique to diagnose osteoporosis and is seldomly used in *in vivo* coating assessment studies [86].

- **MRI (magnetic resonance imaging)** does, in contrast to other imaging techniques, not rely on ionizing radiation but on the magnetic spin of protons. Due to the high water content (and thus protons) of soft tissue, MRI is one of the main imaging techniques to assess the musculoskeletal tissues, like cartilage and tendons. MRI only allows indirect imaging of bony structures due to the limited water content of the bone. The main drawbacks for MRI are the duration of the imaging acquisition and the inability for it to be used in combination with metallic implants [76, 84].

- **PET (positron emission tomography)** is based on the detection of the annihilation event of a positron with an electron (beta-decay). Every annihilation-event results in 2 gamma-photons in an opposite direction from the point of decay. The detections of the photons on the detectorring of the scanner results in a 3D image [97]. ^{18}F is one of the most frequently used isotopes (connected to a carrier molecule) to serve as a PET-tracer in orthopaedic research. ^{18}F-fluorodeoxyglucose (FDG) is used for the detection of infection and inflammation (figure 2) and ^{18}F-fluoride as a tracer for bone remodeling [85, 94, 98]. With signal specificity as its advantage, PET does not provide anatomical information, merely the location of the tracer uptake. This is the main reason why PET and CT are often combined in the clinic.

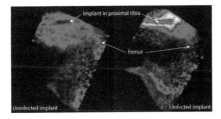

Figure 2. FDG PET of an uninfected implant versus an infected implant in the proximal part of a rabbit tibia, six weeks after surgery. The increased tracer uptake around the infected implant (black area) depicts the local osteomyeliticleasion.

Analytical method / Clinical parameters	Detection method	Detects	Ref.
Body weight	Weighing scale	General physical condition • Weight loss after surgical intervention, returns to pre-operative values within first weeks after intervention. • Persistent weight loss indicates animal discomfort, infection or another systemic event related to device or intervention.	[49, 78, 79, 81, 87]
Temperature	Thermal probe	General physical condition • Post-operative thermal elevation, returns to normal within days after surgery. • Persistent elevation indicates infection or another systemic event related to device or intervention.	[49, 81]
ESR • Erythrocyte sedimentation rate	Anticoagulated blood • Capillary tube	Infection by increase of erythrocyte sedimentation • Elevation in first post-operative week due to surgical intervention. • Remains elevated in case of infection.	[49, 78, 80, 81, 88]
CRP • C-reactive protein	Serum/plasma • ELISA	Infection by increase of CRP levels • Elevation in first post-operative week due to surgical intervention. • Remains elevated in case of infection.	[84]
Leucocyte count and Leucocyte differentiation	Anticoagulated blood • Cell count	Infection by shift in leucocyte distribution • ↑ Neutrophils and monocytes → bacterial infection • ↑ Lymphocytes → viral infection or tumor • ↑ Basophils and eosinophils → inflammatory processes and/or allergic reactions	[49, 65, 78, 80, 81, 84, 88]

Table 4. *In vivo* analytical methods – part 1

Analytical method	Detection method	Detects	Ref.
Imaging modalities			
Optical imaging	Fluorescence or bioluminescence • Fluorescence → emission after excitation • Bioluminescence → auto-emission	Presence of light emitting cells • Fluorescence → GFP • Bioluminescence → luciferase	[76]
X-ray	Electromagnetic radiation	Bone, bone pathology and metal objects	[33, 49, 76, 78-81, 84, 87-94]
CT • Computed tomography	Electromagnetic radiation • X-ray (3D)	Bone, bone pathology and metal objects	[76, 84, 86, 88, 91, 94]
DEXA • Dual energy X-ray absorptiometry	Electromagnetic radiation • Dual energy X-ray of same object	Bone density • Difference in signal between both X-rays allows calculation of bone density	[86]
MRI • Magnetic resonance imaging	Nuclear magnetic resonance • Proton magnetic spin resonance (3D)	Soft tissue • Detection of tissues with a high water content	[76, 84, 88]
Bone scintigraphy	γ-radiation • Single photon emission (2D)	Active bone remodeling • ^{99m}Tc –MDP → increased osteogenesis • ^{67}Ga-citrate → leucocyte activation (e.g. infection)	[84, 88, 95]
SPECT • Single photon emission computed tomography	γ-radiation • Single photon emission (3D)	Active bone remodeling • ^{99m}Tc –MDP → increased osteogenesis • ^{67}Ga-citrate → leucocyte activation (e.g. infection)	[76, 88]
PET • Positron emission tomography	γ-radiation • Positron mediated dual photon emission (3D)	Active bone remodeling and infection • ^{18}F-Fluorodeoxyglucose → inflammation and infection • ^{18}F-Sodiumfluoride → active bone remodeling	[76, 84, 88, 91-94]

Table 4. *In vivo* analytical methods – part 2

Analytical method	Detection method	Detects	Ref.
Other – *Ex vivo*			
Calcium binding fluorophores • Calcein blue • Calcein green • Tetracycline • Xylenol orange • Alizarin red S	Fluorescence (excitation/emission) • Calcein blue (370 / 435 nm) • Calcein green (470 / 530 nm) • Tetracycline (390 / 570 nm) • Xylenol orange (470 / 610 nm) • Alizarin red S (550 / 620 nm)	Calcium deposition during bone remodeling (color) • Calcein blue → blue • Calcein green → green • Tetracycline → yellow • Xylenol orange → orange/red • Alizarin red S → red	[96]
Histology • Paraffin • PMMA	Light microscopy	Tissue specific staining (including bacteria) • Haematoxylin/eosin staining (general) • Masson Goldnertrichrome staining (general) • Gram staining (bacteria) • Wear particles (e.g. polyethylene by polarized light) • Immunostaining (antibody specific)	[33, 49, 75, 79-81, 84, 87, 90-94]
SEM • Scanning electron microscopy	Electron microscopy	Surface treatment, bacterial biofilm, wear particles • Surface assessment of non-metallic substrates by gold or carbon sputtering • Metallic substrates can be assessed directly	[65, 80]
Bacterial/bone culture • Tissue swaps • Bone homogenates	(Selective) culture media • Tryptic soy agar/broth • Tellurite glycine agar (selective for coagulase negative Staphylococci)	Bacterial growth under specific circumstances • Quantification by colony count or OD 600 measurements	[48, 49, 65, 78-81, 84, 87, 89-94]

Table 4. *In vivo* analytical methods – part 3

- **A bone scan** (bone scintigraphy) is based on the direct detection of gamma radiation originating from the injected tracer molecule (often ^{99m}Tc, ^{67}Ga or ^{111}In) connected to a specific ligand which allows tissue specific binding and thus imaging. A bone scan provides two-dimensional images of the patient, which are sufficient in the clinic for the diagnosis [84].

- **SPECT** (single photon emission computed tomography) on the other hand allows acquisition of three-dimensional images, providing more insight in size and localization of certain pathology. In general, bone scan/SPECT-tracers have a longer half-life than PET tracers making them more cost-effective to produce. Just like PET, SPECT provides limited anatomical information and is therefore often combined with CT in the clinic [76].

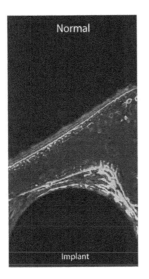 

Figure 3. The use of calcium binding fluorophores, depicted in 50 micron PMMA sections, of a rabbit tibial intrame-dullary nail model, to address normal bone remodeling and bone remodeling in case of an implant infection. Calcein green was injected at 2 weeks, xylenol orange at 4 weeks and calcein blue at 6 weeks. In the case of normal bone remodeling, calcium deposition is detected around the implant, combined with bone remodeling of the cortical wall. In case of an implant infection the most calcium deposition is located in the outer cortical wall depicting the periosteal elevation and calcification during the 6 week follow-up.

4.2.3. Ex vivo analysis

- **Calcium binding fluorophores** (like calcein green, blue and xylenol orange) are being used for the *in vivo* labeling of the calcium deposition at the time of injection. The use of different fluorophores, emitting at different wavelengths, allow post-mortem visualization of the calcium deposition during the experimental follow-up [96]. This provides the opportunity to determine implant ingrowth and bone remodeling in a normal healthy situation and periosteal elevation and calcification during the progression of an osteomyelitis (Figure 3).

- **Histology** is a commonly used method to assess the tissue in the implant area on e.g. tissue morphology or bacterial presence. Tissue sections with metallic implants generally require embedding in methylmethacrylate, instead of paraffin, with the inability to allow immunostainings as a major drawback. Still it provides the unique opportunity to assess the tissue-implant interface [33, 49, 57, 80].

- **Electronmicroscopy** allows analysis of the implant surface (with or without coating) after distraction from the surrounding tissue. This can include analysis of the bone matrix-implant interface but also analysis of a formed biofilm [31].

5. Conclusion

Both osteointegration and infections are of concern in implants and prosthesis used in the field of orthopaedic and trauma surgery. Metallic alloys used for plating and nailing of fractures and joint replacements are the largest group of these implants. Hydroxyapatite coatings have proven to be successful to promote osseous integration of uncemented total hip prosthesis. During the last years the focus on coating development has shifted from osteoconductive coatings (like hydroxyapatite) towards osteoinductive coatings to support bone remodeling (like RGD and BMP coatings) and antimicrobial coatings for implant infection treatment and prophylaxis (like silver or antibiotic releasing coatings).

Plasma spraying is the most used and accepted method for hydroxyapatite coatings. Other coating techniques which do not require high temperatures are necessary for the application of bioactive coatings that promote osteogenesis and/or prevent infections.

With the current palette of *in vitro* (e.g. MTT, ALP and SEM), *in vivo* (e.g. ESR, CT and PET) and *ex vivo* techniques (e.g. bacterial culture, calcium binding fluorophores and histology), we can thoroughly evaluate novel implant coatings in a qualitative and quantitative fashion. The strength of such an evaluation will always lie in the combination of the individual methods, leading to a complete, broad-spectrum analysis on coating toxicity and efficacy.

Author details

Jim C.E. Odekerken, Tim J.M. Welting, Jacobus J.C. Arts, Geert H.I.M. Walenkamp and Pieter J. Emans*

*Address all correspondence to: p.emans@mumc.nl

Department of Orthopaedic Surgery, Research school CAPHRI, Maastricht University Medical Centre, the Netherlands

References

[1] Adams, J. E. A Simple Method of Mechanical Fixation for Fracture of Long Bones. Br Med J. (1918). Jan 5;, 1(2975), 12-3.

[2] Devas, M. B. Arthroplasty of the hip: a review of 110 cup and replacement arthroplasties. The Journal of bone and joint surgery British (1954). Nov;36-B(4):561-6., 1954

[3] Judet, J, & Judet, R. The use of an artificial femoral head for arthroplasty of the hip joint. The Journal of bone and joint surgery British (1950). May;32-B(2):166-73., 1950

[4] Surveillance Report: Annual epidemiological reportreporting on 2009 surveillance data and 2010 epidemic intelligence data. http://ecdc.europa.eu/en/publications/publications/0910_sur_annual_epidemiological_report_on_communicable_diseases_in_europe.pdfAccessed December 2012): European Centre for Disease prevention and Control; (2011).

[5] Dale, H, Hallan, G, Espehaug, B, Havelin, L. I, & Engesaeter, L. B. Increasing risk of revision due to deep infection after hip arthroplasty. Acta orthopaedica. [Comparative Study]. (2009). Dec;, 80(6), 639-45.

[6] Healthcare-associated infections- Fact sheethttp://www.who.int/gpsc/country_work/gpsc_ccisc_fact_sheet_en.pdfAccessed December 2012): World Health organisation; (2012).

[7] Surveillance of healthcare-associated infections in Europehttp://www.ecdc.europa.eu/en/publications/Publications/120215_SUR_HAI_2007.pdfAccessed December 2012): European Centre of Disease prevention and Control; (2007).

[8] Capello, W. N, Antonio, D, Jaffe, J. A, Geesink, W. L, Manley, R. G, Feinberg, M. T, & Hydroxyapatite-coated, J. R. femoral components: 15-year minimum followup. Clinical orthopaedics and related research. (2006). Dec;, 453, 75-80.

[9] Tannast, M, Najibi, S, & Matta, J. M. Two to twenty-year survivorship of the hip in 810 patients with operatively treated acetabular fractures. The Journal of bone and joint surgery American volume. [Comparative Study, Evaluation Studies, Research Support, Non-U.S. Gov't]. (2012). Sep 5;, 94(17), 1559-67.

[10] Gristina, A. G. Biomaterial-centered infection: microbial adhesion versus tissue integration. Science. [Research Support, U.S. Gov't, P.H.S.]. (1987). Sep 25;, 237(4822), 1588-95.

[11] Busscher, H. J, Van Der Mei, H. C, Subbiahdoss, G, & Jutte, P. C. van den Dungen JJ, Zaat SA, et al. Biomaterial-associated infection: locating the finish line in the race for the surface. Sci Transl Med. (2012). Sep 26;4(153):153rv10.

[12] Arciola, C. R, Campoccia, D, Speziale, P, Montanaro, L, & Costerton, J. W. Biofilm formation in Staphylococcus implant infections. A review of molecular mechanisms

and implications for biofilm-resistant materials. Biomaterials. [Review]. (2012). Sep;, 33(26), 5967-82.

[13] Geurts, J. Chris Arts JJ, Walenkamp GH. Bone graft substitutes in active or suspected infection. Contra-indicated or not? Injury. [Review]. (2011). Sep;42 Suppl 2:S, 82-6.

[14] Geesink, R. G. Osteoconductive coatings for total joint arthroplasty. Clinical orthopaedics and related research. [Research Support, Non-U.S. Gov't, Review]. (2002). Feb(395):53-65.

[15] Geesink, R. G, De Groot, K, & Klein, C. P. Chemical implant fixation using hydroxylapatite coatings. The development of a human total hip prosthesis for chemical fixation to bone using hydroxyl-apatite coatings on titanium substrates. Clinical orthopaedics and related research. (1987). Dec(225):147-70.

[16] Geesink, R. G, De Groot, K, & Klein, C. P. Bonding of bone to apatite-coated implants. The Journal of bone and joint surgery British (1988). Jan;70(1):17-22., 1988

[17] Smabrekke, A, Espehaug, B, Havelin, L. I, & Furnes, O. Operating time and survival of primary total hip replacements: an analysis of 31,745 primary cemented and uncemented total hip replacements from local hospitals reported to the Norwegian Arthroplasty Register 1987-2001. Acta Orthop Scand. [Comparative Study]. (2004). Oct;, 75(5), 524-32.

[18] LeGeros RZProperties of osteoconductive biomaterials: calcium phosphates. Clinical orthopaedics and related research. [Research Support, Non-U.S. Gov't, Research Support, U.S. Gov't, P.H.S., Review]. (2002). Feb(395):81-98.

[19] Jarcho, M. Calcium phosphate ceramics as hard tissue prosthetics. Clinical orthopaedics and related research. [Review]. (1981). Jun(157):259-78.

[20] Geesink, R. G. Hydroxyapatite-coated total hip prostheses. Two-year clinical and roentgenographic results of 100 cases. Clinical orthopaedics and related research. (1990). Dec(261):39-58.

[21] Charles, L. F, Shaw, M. T, Olson, J. R, & Wei, M. Fabrication and mechanical properties of PLLA/PCL/HA composites via a biomimetic, dip coating, and hot compression procedure. Journal of materials science Materials in medicine. [Research Support, U.S. Gov't, Non-P.H.S.]. (2010). Jun;, 21(6), 1845-54.

[22] Ding, S. J. Properties and immersion behavior of magnetron-sputtered multi-layered hydroxyapatite/titanium composite coatings. Biomaterials. [Research Support, Non-U.S. Gov't]. (2003). Oct;, 24(23), 4233-8.

[23] Saju, K. K, Reshmi, R, Jayadas, N. H, James, J, & Jayaraj, M. K. Polycrystalline coating of hydroxyapatite on TiAl6implant material grown at lower substrate temperatures by hydrothermal annealing after pulsed laser deposition. Proc Inst Mech Eng H. [Evaluation Studies]. (2009). Nov;223(8):1049-57., 4

[24] Zhao, J, Xiao, S, Lu, X, Wang, J, & Weng, J. A study on improving mechanical proper-
      ties of porous HA tissue engineering scaffolds by hot isostatic pressing. Biomed Ma-
      ter. [Evaluation Studies, Research Support, Non-U.S. Gov't]. (2006). Dec;, 1(4), 188-92.

[25] Boccaccini, A. R, Keim, S, Ma, R, Li, Y, & Zhitomirsky, I. Electrophoretic deposition
      of biomaterials. J R Soc Interface. [Review]. (2010). Oct 6;7 Suppl 5:S, 581-613.

[26] Leeuwenburgh, S, Wolke, J, Schoonman, J, & Jansen, J. Electrostatic spray deposition
      (ESD) of calcium phosphate coatings. Journal of biomedical materials research Part
      A. (2003). Aug 1;, 66(2), 330-4.

[27] Leeuwenburgh, S. C, Wolke, J. G, Siebers, M. C, Schoonman, J, & Jansen, J. A. In vitro
      and in vivo reactivity of porous, electrosprayed calcium phosphate coatings. Bioma-
      terials. [Research Support, Non-U.S. Gov't]. (2006). Jun;, 27(18), 3368-78.

[28] Heimann, R. B. Thermal spraying of biomaterials. Surface & Coatings Technology.
      2006 (2006).

[29] Kim, H. W, Knowles, J. C, Salih, V, & Kim, H. E. Hydroxyapatite and fluor-hydrox-
      yapatite layered film on titanium processed by a sol-gel route for hard-tissue im-
      plants. J Biomed Mater Res B Appl Biomater. (2004). Oct 15;, 71(1), 66-76.

[30] Vidigal, G. M. Jr., Groisman M, de Sena LA, Soares Gde A. Surface characterization
      of dental implants coated with hydroxyapatite by plasma spray and biomimetic
      process. Implant Dent. [In Vitro, Research Support, Non-U.S. Gov't]. (2009). Aug;,
      18(4), 353-61.

[31] Leeuwenburgh, S. C, Wolke, J. G, Lommen, L, Pooters, T, Schoonman, J, & Jansen, J.
      A. Mechanical properties of porous, electrosprayed calcium phosphate coatings.
      Journal of biomedical materials research Part A. [Research Support, Non-U.S. Gov't].
      (2006). Sep 1;, 78(3), 558-69.

[32] Flynn, O, & Stanton, K. P. KT. Optimisation of the enamelling of an apatite-mullite
      glass-ceramic coating on Ti6Al4V. Journal of materials science Materials in medicine.
      [Research Support, Non-U.S. Gov't]. (2011). Sep;, 22(9), 2035-44.

[33] Alt, V, Bitschnau, A, Bohner, F, Heerich, K. E, Magesin, E, Sewing, A, et al. Effects of
      gentamicin and gentamicin-RGD coatings on bone ingrowth and biocompatibility of
      cementless joint prostheses: an experimental study in rabbits. Acta Biomater. [In Vi-
      tro, Research Support, Non-U.S. Gov't]. (2011). Mar;, 7(3), 1274-80.

[34] Liu, Y, Hunziker, E. B, Layrolle, P, De Bruijn, J. D, & De Groot, K. Bone morphoge-
      netic protein 2 incorporated into biomimetic coatings retains its biological activity.
      Tissue Eng. (2004). Jan-Feb;10(1-2):101-8.

[35] Ardjomandi, N, Klein, C, Kohler, K, Maurer, A, Kalbacher, H, Niederlander, J, et al.
      Indirect coating of RGD peptides using a poly-L-lysine spacer enhances jaw perios-
      teal cell adhesion, proliferation, and differentiation into osteogenic tissue. Journal of

biomedical materials research Part A. [Research Support, Non-U.S. Gov't]. (2012). Aug;, 100(8), 2034-44.

[36]   Wang, Y. J, Lin, F. H, Sun, J. S, Huang, Y. C, Chueh, S. C, & Hsu, F. Y. Collagen-hydroxyapatite microspheres as carriers for bone morphogenic protein-4. Artif Organs. (2003). Feb;, 27(2), 162-8.

[37]   Sun, J. S, Lin, F. H, Wang, Y. J, Huang, Y. C, Chueh, S. C, & Hsu, F. Y. Collagen-hydroxyapatite/tricalcium phosphate microspheres as a delivery system for recombinant human transforming growth factor-beta 1. Artif Organs. (2003). Jul;, 27(7), 605-12.

[38]   Haidar, Z. S, Tabrizian, M, & Hamdy, R. C. A hybrid rhOP-1 delivery system enhances new bone regeneration and consolidation in a rabbit model of distraction osteogenesis. Growth Factors. [Research Support, Non-U.S. Gov't]. (2010). Feb;, 28(1), 44-55.

[39]   Knippenberg, M, & Helder, M. N. Zandieh Doulabi B, Wuisman PI, Klein-Nulend J. Osteogenesis versus chondrogenesis by BMP-2 and BMP-7 in adipose stem cells. Biochem Biophys Res Commun. [Research Support, Non-U.S. Gov't]. (2006). Apr 14;, 342(3), 902-8.

[40]   Liu, Y, Wu, G, & De Groot, K. Biomimetic coatings for bone tissue engineering of critical-sized defects. J R Soc Interface. [Review]. (2010). Oct 6;7 Suppl 5:S, 631-47.

[41]   Liu, Y, De Groot, K, & Hunziker, E. B. BMP-2 liberated from biomimetic implant coatings induces and sustains direct ossification in an ectopic rat model. Bone. [Research Support, Non-U.S. Gov't]. (2005). May;, 36(5), 745-57.

[42]   Crockarell, J. R, Hanssen, A. D, Osmon, D. R, & Morrey, B. F. Treatment of infection with debridement and retention of the components following hip arthroplasty. The Journal of bone and joint surgery American (1998). Sep;80(9):1306-13., 1998

[43]   Kim, Y. H, Kim, J. S, Park, J. W, & Joo, J. H. Cementless revision for infected total hip replacements. The Journal of bone and joint surgery British volume. [Evaluation Studies]. (2011). Jan;, 93(1), 19-26.

[44]   Walenkamp, G. H. Gentamicin PMMA beads and other local antibiotic carriers in two-stage revision of total knee infection: a review. J Chemother. [Review]. (2001). Nov;13 Spec No , 1(1), 66-72.

[45]   Walenkamp, G. H, Kleijn, L. L, & De Leeuw, M. Osteomyelitis treated with gentamicin-PMMA beads: 100 patients followed for 1-12 years. Acta Orthop Scand. (1998). Oct;, 69(5), 518-22.

[46]   Rasyid, H. N, Van Der Mei, H. C, Frijlink, H. W, Soegijoko, S, Van Horn, J. R, Busscher, H. J, et al. Concepts for increasing gentamicin release from handmade bone cement beads. Acta orthopaedica. (2009). Oct;, 80(5), 508-13.

[47] Kalicke, T, Schierholz, J, Schlegel, U, Frangen, T. M, Koller, M, Printzen, G, et al. Effect on infection resistance of a local antiseptic and antibiotic coating on osteosynthesis implants: an in vitro and in vivo study. Journal of orthopaedic research : official publication of the Orthopaedic Research Society. [In Vitro, Research Support, Non-U.S. Gov't]. (2006). Aug;, 24(8), 1622-40.

[48] Darouiche, R. O, Mansouri, M. D, Zakarevicz, D, Alsharif, A, & Landon, G. C. In vivo efficacy of antimicrobial-coated devices. J Bone Joint Surg Am. (2007). Apr;, 89(4), 792-7.

[49] Moojen, D. J, Vogely, H. C, Fleer, A, Nikkels, P. G, Higham, P. A, Verbout, A. J, et al. Prophylaxis of infection and effects on osseointegration using a tobramycin-periapatite coating on titanium implants--an experimental study in the rabbit. Journal of orthopaedic research : official publication of the Orthopaedic Research Society. [Research Support, Non-U.S. Gov't]. (2009). Jun;, 27(6), 710-6.

[50] Neut, D, Kluin, O. S, Crielaard, B. J, Van Der Mei, H. C, Busscher, H. J, & Grijpma, D. W. A biodegradable antibiotic delivery system based on poly-(trimethylene carbonate) for the treatment of osteomyelitis. Acta orthopaedica. [Comparative Study]. (2009). Oct;, 80(5), 514-9.

[51] Fuchs, T, Stange, R, Schmidmaier, G, & Raschke, M. J. The use of gentamicin-coated nails in the tibia: preliminary results of a prospective study. Archives of orthopaedic and trauma surgery. (2011). Oct;, 131(10), 1419-25.

[52] Darouiche, R. O. Anti-infective efficacy of silver-coated medical prostheses. Clinical infectious diseases : an official publication of the Infectious Diseases Society of America. [Review]. (1999). Dec;quiz 8., 29(6), 1371-7.

[53] Petering, H. G. Pharmacology and toxicology of heavy metals: Silver. Pharmacology & Therapeutics Part A: Chemotherapy, Toxicology and Metabolic Inhibitors. (1976).

[54] Shimazaki, T, Miyamoto, H, Ando, Y, Noda, I, Yonekura, Y, Kawano, S, et al. In vivo antibacterial and silver-releasing properties of novel thermal sprayed silver-containing hydroxyapatite coating. Journal of biomedical materials research Part B, Applied biomaterials. (2010). Feb;, 92(2), 386-9.

[55] Chen, W, Liu, Y, Courtney, H. S, Bettenga, M, Agrawal, C. M, Bumgardner, J. D, et al. In vitro anti-bacterial and biological properties of magnetron co-sputtered silver-containing hydroxyapatite coating. Biomaterials. [Research Support, N.I.H., Extramural]. (2006). Nov;, 27(32), 5512-7.

[56] Moseke, C, Gbureck, U, Elter, P, Drechsler, P, Zoll, A, Thull, R, et al. Hard implant coatings with antimicrobial properties. Journal of materials science Materials in medicine. [Research Support, Non-U.S. Gov't]. (2011). Dec;, 22(12), 2711-20.

[57] Badiou, W, Lavigne, J. P, Bousquet, P. J, Callaghan, O, Mares, D, & De Tayrac, P. R. In vitro and in vivo assessment of silver-coated polypropylene mesh to prevent infec-

tion in a rat model. Int Urogynecol J. [Evaluation Studies, In Vitro]. (2011). Mar;, 22(3), 265-72.

[58] Moojen, D. J, Vogely, H. C, Fleer, A, Verbout, A. J, Castelein, R. M, & Dhert, W. J. No efficacy of silver bone cement in the prevention of methicillin-sensitive Staphylococ-cal infections in a rabbit contaminated implant bed model. Journal of orthopaedic re-search : official publication of the Orthopaedic Research Society. [Research Support, Non-U.S. Gov't]. (2009). Aug;, 27(8), 1002-7.

[59] Coester, L. M, Nepola, J. V, Allen, J, & Marsh, J. L. The effects of silver coated exter-nal fixation pins. Iowa Orthop J. [Randomized Controlled Trial]. (2006). , 26, 48-53.

[60] Masse, A, Bruno, A, Bosetti, M, Biasibetti, A, Cannas, M, & Gallinaro, P. Prevention of pin track infection in external fixation with silver coated pins: clinical and micro-biological results. J Biomed Mater Res. [Controlled Clinical Trial]. (2000). Sep;, 53(5), 600-4.

[61] Pickard, R, Lam, T, Maclennan, G, Starr, K, Kilonzo, M, Mcpherson, G, et al. Antimi-crobial catheters for reduction of symptomatic urinary tract infection in adults re-quiring short-term catheterisation in hospital: a multicentre randomised controlled trial. Lancet. (2012). Nov 2.

[62] Evans, N. D, Gentleman, E, Chen, X, Roberts, C. J, Polak, J. M, & Stevens, M. M. Ex-tracellular matrix-mediated osteogenic differentiation of murine embryonic stem cells. Biomaterials. [Research Support, Non-U.S. Gov't]. (2010). Apr;, 31(12), 3244-52.

[63] Jung, G. Y, Park, Y. J, & Han, J. S. Effects of HA released calcium ion on osteoblast differentiation. Journal of materials science Materials in medicine. [Research Support, Non-U.S. Gov't]. (2010). May;, 21(5), 1649-54.

[64] Kapoor, R, Sistla, P. G, Kumar, J. M, Raj, T. A, Srinivas, G, Chakraborty, J, et al. Com-parative assessment of structural and biological properties of biomimetically coated hydroxyapatite on alumina (alpha-Al2O3) and titanium (Ti-6Al-4V) alloy substrates. Journal of biomedical materials research Part A. [Evaluation Studies, Research Sup-port, Non-U.S. Gov't]. (2010). Sep 1;, 94(3), 913-26.

[65] Moskowitz, J. S, Blaisse, M. R, Samuel, R. E, Hsu, H. P, Harris, M. B, Martin, S. D, et al. The effectiveness of the controlled release of gentamicin from polyelectrolyte mul-tilayers in the treatment of Staphylococcus aureus infection in a rabbit bone model. Biomaterials. [Research Support, N.I.H., Extramural]. (2010). Aug;, 31(23), 6019-30.

[66] Qiang, F, Rahaman, M. N, Nai, Z, Wenhai, H, Deping, W, Liying, Z, et al. In vitro study on different cell response to spherical hydroxyapatite nanoparticles. J Biomater Appl. (2008). Jul;, 23(1), 37-50.

[67] De Carlos, A, Lusquinos, F, Pou, J, Leon, B, Perez-amor, M, Driessens, F. C, et al. In vitro testing of Nd:YAG laser processed calcium phosphate coatings. Journal of ma-

terials science Materials in medicine. [Comparative Study, Research Support, Non-U.S. Gov't]. (2006). Nov;, 17(11), 1153-60.

[68] Kazemzadeh-narbat, M, Noordin, S, Masri, B. A, Garbuz, D. S, Duncan, C. P, Hancock, R. E, et al. Drug release and bone growth studies of antimicrobial peptide-loaded calcium phosphate coating on titanium. Journal of biomedical materials research Part B, Applied biomaterials. [Research Support, Non-U.S. Gov't]. (2012). Jul;, 100(5), 1344-52.

[69] Hu, H, Zhang, W, Qiao, Y, Jiang, X, Liu, X, & Ding, C. Antibacterial activity and increased bone marrow stem cell functions of Zn-incorporated TiO2 coatings on titanium. Acta Biomater. [Research Support, Non-U.S. Gov't]. (2012). Feb;, 8(2), 904-15.

[70] Khalilpour, P, Lampe, K, Wagener, M, Stigler, B, Heiss, C, Ullrich, M. S, et al. Ag/SiO(x)C(y) plasma polymer coating for antimicrobial protection of fracture fixation devices. Journal of biomedical materials research Part B, Applied biomaterials. [Evaluation Studies]. (2010). Jul;, 94(1), 196-202.

[71] Lanfer, B, Seib, F. P, Freudenberg, U, Stamov, D, Bley, T, Bornhauser, M, et al. The growth and differentiation of mesenchymal stem and progenitor cells cultured on aligned collagen matrices. Biomaterials. (2009). Oct;, 30(30), 5950-8.

[72] Vichai, V, & Kirtikara, K. Sulforhodamine B colorimetric assay for cytotoxicity screening. Nat Protoc. (2006). , 1(3), 1112-6.

[73] Necula, B. S, Van Leeuwen, J. P, Fratila-apachitei, L. E, Zaat, S. A, Apachitei, I, & Duszczyk, J. In vitro cytotoxicity evaluation of porous TiO(2)-Ag antibacterial coatings for human fetal osteoblasts. Acta Biomater. (2012). Nov;, 8(11), 4191-7.

[74] Goransson, A, Arvidsson, A, Currie, F, Franke-stenport, V, Kjellin, P, Mustafa, K, et al. An in vitro comparison of possibly bioactive titanium implant surfaces. J Biomed Mater Res A. (2009). Mar 15;, 88(4), 1037-47.

[75] Woods, G. L, & Walker, D. H. Detection of infection or infectious agents by use of cytologic and histologic stains. Clin Microbiol Rev. [Review]. (1996). Jul;, 9(3), 382-404.

[76] Tremoleda, J. L, Khalil, M, Gompels, L. L, Wylezinska-arridge, M, Vincent, T, & Gsell, W. Imaging technologies for preclinical models of bone and joint disorders. EJNMMI Res. (2011).

[77] Rodet, A. Etude expérimentale sur l'osteomyelite infectieuse. Compte rend. Acad de Science. (1884).

[78] Norden, C. W. Experimental osteomyelitis. I. A description of the model. J Infect Dis. (1970). Nov;, 122(5), 410-8.

[79] Andriole, V. T, Nagel, D. A, & Southwick, W. O. A paradigm for human chronic osteomyelitis. The Journal of bone and joint surgery American (1973). Oct;55(7):1511-5., 1973

[80] Vogely, H. C, Oosterbos, C. J, Puts, E. W, Nijhof, M. W, Nikkels, P. G, Fleer, A, et al. Effects of hydrosyapatite coating on Ti-6A1-4V implant-site infection in a rabbit tibial model. Journal of orthopaedic research : official publication of the Orthopaedic Research Society. (2000). May;, 18(3), 485-93.

[81] Poultsides, L. A, Papatheodorou, L. K, Karachalios, T. S, Khaldi, L, Maniatis, A, Petinaki, E, et al. Novel model for studying hematogenous infection in an experimental setting of implant-related infection by a community-acquired methicillin-resistant S. aureus strain. J Orthop Res. (2008). Oct;, 26(10), 1355-62.

[82] Shih, L. Y, Wu, J. J, & Yang, D. J. Erythrocyte sedimentation rate and C-reactive protein values in patients with total hip arthroplasty. Clinical orthopaedics and related research. [Case Reports]. (1987). Dec(225):238-46.

[83] Calhoun, J. H, Manring, M. M, & Shirtliff, M. Osteomyelitis of the long bones. Semin Plast Surg. (2009). May;, 23(2), 59-72.

[84] Zimmerli, W. Infection and musculoskeletal conditions: Prosthetic-joint-associated infections. Best Pract Res Clin Rheumatol. (2006). Dec;, 20(6), 1045-63.

[85] Monjo, M, Lamolle, S. F, Lyngstadaas, S. P, Ronold, H. J, & Ellingsen, J. E. In vivo expression of osteogenic markers and bone mineral density at the surface of fluoride-modified titanium implants. Biomaterials. (2008). Oct;, 29(28), 3771-80.

[86] MacNeil JABoyd SK. Accuracy of high-resolution peripheral quantitative computed tomography for measurement of bone quality. Med Eng Phys. (2007). Dec;, 29(10), 1096-105.

[87] Sanzen, L, & Linder, L. Infection adjacent to titanium and bone cement implants: an experimental study in rabbits. Biomaterials. [Research Support, Non-U.S. Gov't]. (1995). Nov;, 16(16), 1273-7.

[88] El-Maghraby, T. A, Moustafa, H. M, & Pauwels, E. K. Nuclear medicine methods for evaluation of skeletal infection among other diagnostic modalities. Q J Nucl Med Mol Imaging. [Review]. (2006). Sep;, 50(3), 167-92.

[89] Melcher, G. A, Claudi, B, Schlegel, U, Perren, S. M, Printzen, G, & Munzinger, J. Influence of type of medullary nail on the development of local infection. An experimental study of solid and slotted nails in rabbits. The Journal of bone and joint surgery British volume. [Comparative Study, Research Support, Non-U.S. Gov't]. (1994). Nov;, 76(6), 955-9.

[90] An, Y. H, Bradley, J, Powers, D. L, & Friedman, R. J. The prevention of prosthetic infection using a cross-linked albumin coating in a rabbit model. The Journal of bone

and joint surgery British volume. [Research Support, Non-U.S. Gov't]. (1997). Sep;, 79(5), 816-9.

[91] Koort, J. K, Makinen, T. J, Knuuti, J, Jalava, J, & Aro, H. T. Comparative 18F-FDG PET of experimental Staphylococcus aureus osteomyelitis and normal bone healing. J Nucl Med. (2004). Aug;, 45(8), 1406-11.

[92] Jones-jackson, L, Walker, R, Purnell, G, Mclaren, S. G, Skinner, R. A, Thomas, J. R, et al. Early detection of bone infection and differentiation from post-surgical inflammation using 2-deoxy-2-[18F]-fluoro-D-glucose positron emission tomography (FDG-PET) in an animal model. J Orthop Res. (2005). Nov;, 23(6), 1484-9.

[93] Makinen, T. J, Veiranto, M, Knuuti, J, Jalava, J, Tormala, P, & Aro, H. T. Efficacy of bioabsorbable antibiotic containing bone screw in the prevention of biomaterial-related infection due to Staphylococcus aureus. Bone. [Comparative Study, Research Support, Non-U.S. Gov't]. (2005). Feb;, 36(2), 292-9.

[94] Lankinen, P, Lehtimaki, K, Hakanen, A. J, Roivainen, A, & Aro, H. T. A comparative 18 F-FDG PET/CT imaging of experimental Staphylococcus aureus osteomyelitis and Staphylococcus epidermidis foreign-bodyassociated infection in the rabbit tibia. EJNMMI Res. (2012). Jul 23;2(1):41.

[95] Guhlmann, A, Brecht-krauss, D, Suger, G, Glatting, G, Kotzerke, J, Kinzl, L, et al. Fluorine-18-FDG PET and technetium-99m antigranulocyte antibody scintigraphy in chronic osteomyelitis. Journal of nuclear medicine : official publication, Society of Nuclear Medicine. (1998). Dec;, 39(12), 2145-52.

[96] Van Gaalen, S. M, Kruyt, M. C, Geuze, R. E, De Bruijn, J. D, Alblas, J, & Dhert, W. J. Use of fluorochrome labels in in vivo bone tissue engineering research. Tissue Eng Part B Rev. [Review]. (2010). Apr;, 16(2), 209-17.

[97] Schlyer, D. J. PET tracers and radiochemistry. Ann Acad Med Singapore. (2004). Mar;, 33(2), 146-54.

[98] Segall, G, Delbeke, D, Stabin, M. G, Even-sapir, E, Fair, J, Sajdak, R, et al. SNM practice guideline for sodium 18F-fluoride PET/CT bone scans 1.0. Journal of nuclear medicine : official publication, Society of Nuclear Medicine. [Practice Guideline]. (2010). Nov;, 51(11), 1813-20.

Chapter 4

Nanocoatings

R. Abdel-Karim and A. F. Waheed

Additional information is available at the end of the chapter

1. Introduction

Nanocoatings are one of the most important topics within the range of nanotechnology. Through nanoscale engineering of surfaces and layers, a vast range of functionalities and new physical effects can be achieved. Some application ranges of nanolayers and coatings are summarized in table 1 [1].

Surface Properties	Application Examples
Mechanical proeprties (e.g tribology, hardness, scratch resistance	Wear protection of machinery and equipment, mechanical protection of soft materials. (polymers, wod, textile, etc.), superplasticity of ceramics
Wetting properties (e.g antiadhesive, hydrophobic, hydrophibic)	Antigraffiti, Antifouling, Lotus-effect, self-cleaning surfaces.
Thermal and chemical proeprties(e.g heat resistance and insulation, corrosion resistance)	Corrosion protection for machinery and equipment, heat resistance for turbines and engines, thermal insulation equipment and building materials.
Biological properties (biocompatibility, anti-infective)	Biocompatible implants, medical tools, wound dressings.
Electronical and magnetic properties (e.g magneto resistance, dielectric)	Ultrathin dieelectrics for field effect transistors, magnetoresistive sensors and data memory.
Catalytic efficiency	Better catalytic efficiency through higher surface-to-volume.
Optical properties (e.g anti-reflection, photo-and electrochromatic)	Photo-and electrochromatic windows, antireflective screens and solar cells.

Table 1. Some applications of nanocoatings

Many synthesis techniques for production of nanostructured coatings have been developed such as sputtering, laser ablation, sol/gel technique, chemical vapour deposition, gas-condensation, plasma spraying, and electrochemical deposition [1]. Chemical vapour deposition includes chemical reaction of input materials in the gas phase and deposition of the product on the surface. Physical vapour deposition (PVD) includes transforming the material into the gaseous phase and then deposition on the surface [2]. The impact of an atom or ion on a surface produces sputtering from the surface. Unlike many other vapour phase techniques there is no melting of the material. Sputtering is done at low pressure on cold substrate. In laser ablation, pulsed light from an excimer laser is focused onto a solid target in vacuum to boil off a plum of energetic atom. A substrate will receive a thin film of the target material. The sol-gel process is well adapted for ceramics and composites at room temperature [1].

The superiority of electrochemical deposition techniques in synthesizing various nanomaterials that exhibit improved compared with materials produced by conventional techniques, will be discussed. Nanocoatings can be obtained either directly on substrates or by using porous templates.

2. Electro deposition of nanomaterials

Nano crystalline materials were first reported by Gleiter [3] and due to their attractive properties. The fact that electrochemical deposition "ED", also being an atomic deposition process, can be used to synthesize nanomaterials has generated a great deal of interest in recent years. The obvious advantages of this century-old process of ED are rapidity, low cost, high purity, production of free-standing parts with complex shapes, higher deposition rates, the production of coatings on widely differing substrates. In addition, ability to produce structural features with sizes ranging from nm to µm, and ability to produce compositions unattainable by other techniques [4, 5]. This method also provides for cost-effective production of free-standing forms such as ultrathin foil, wire, sheet, and plate, as well as complex shapes.

Electro deposition parameters are bath composition, pH, temperature, over potential, bath additives, etc... Important microstructural features of the substrate include grain size, crystallographic texture, dislocation density, internal stress. Crystallization (Figure. 1) occurs either by the buildup of existing crystals or the formation of new ones. These two processes are in competition with each other and are influenced by different factors. The two key mechanisms which have been identified as the major rate-determining steps for nanocrystal formation are charge transfer at the electrode surface and surface diffusion of adions on the crystal surface [6]. With increasing inhibition, the deposit structure changes from basis oriented and reproduction type (BR) to twin transition types (TT), to field oriented type (FT), and finally to unoriented dispersion type (UD). A large number of grain refiners have been described in the literature; their effectiveness depends upon surface adsorption characteristics, compatibility with the electrolyte, temperature stability, etc. For example, saccharin, coumarin, thiorea, and HCOOH have all been successfully applied to achieve grain refinement down to the nanocrystalline range for nickel electrodeposits. The second important factor in nanocrystal formation during electro crystallization is overpotential; Grain growth is favored at low over potential and high surface diffusion rates. On the other hand, high over potential and low diffusion rates promote the formation of new nuclei [7].

2.1. Electrodeposited nanocrystalline Ni-Fe alloys

Nanostructured Ni-Fe alloys, produced by electro-deposition technique provide material with significant improved strength and good magnetic properties, without compromising the coefficient of thermal expansion (CTE). Such properties made these alloys to be used in many of the applications where conventional materials are currently used. For such applications a special attention has been made to study the physical, mechanical and chemical properties of such alloys because of the potential for performance enhancement for various applications of Ni-Fe alloys arising from the enhanced properties due to the ultra-fine grain size of these alloys [7-9].

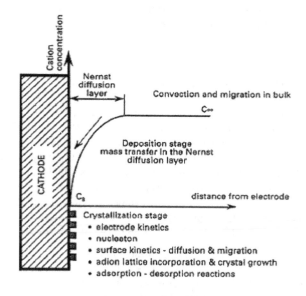

Figure 1. Two stages of electro crystallization according to Bockris et al. [6]

According to R. Abdel-Karim et al. [10], nanocrystalline Ni-Fe deposits with different compo-sition and grain sizes were fabricated by electrodeposition. Deposits with iron contents in the range from 7 to 31% were obtained by changing the Ni^{2+}/Fe^{2+} mass ratio in the electrolyte. The deposits were found to be nanocrystalline with average grain size in the range 20–30 nm. The surface morphology was found to be dependent on Ni^{2+}/Fe^{2+} mass ratio as well as electroplating time. Figure 2 represents SEM of electrodeposited Ni base layers at longer electrode position time (100 min) as a function of Ni^{2+}/Fe^{2+} mass ratio in the electrolytic bath. From Figure 2(a), in case of Ni^{2+}/Fe^{2+} mass ratio equal to 20.7, SEM image displayed well defined nodular coarse and fine particles with no appearance of grain boundaries. This nanosized particles can be better illustrated by using higher magnification (100000x), as shown in Figure 2(b). From Figure 2(c), sample of Ni^{2+}/Fe^{2+} mass ratio equal to 13.8 displayed clusters of fine particles embedded in elongated elliptical ones and some grain boundaries can be seen. By raising the iron content and thus decreasing the Ni^{2+}/Fe^{2+} mass ratio down to 9.8 (Figure 2(d)), the surface morphology showed rough cauliflower structure. The cauliflower morphology particle is made of coagulate particle distributed all over the surface with a flattened grains. The grains size decreased with increasing the iron content, especially in case of short time electroplating. Increasing the electroplating time had no significant effect on grain size. The microhardness of the materials followed the regular Hall-Petch relationship with a maximum value (762 Hv) when applying Ni^{2+}/Fe^{2+} mass ratio equal to 9.8.

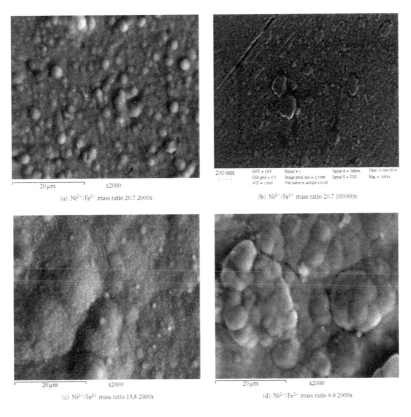

(a) Ni^{2+}/Fe^{2+} mass ratio 20.7 2000x

(b) Ni^{2+}/Fe^{2+} mass ratio 20.7 100000x

(c) Ni^{2+}/Fe^{2+} mass ratio 13,8 2000x

(d) Ni^{2+}/Fe^{2+} mass ratio 9.8 2000x

Figure 2. SEM of electrodeposited Ni-Fe layers at current density 20 mV/cm^2 and deposition time 100 min. as a function of Ni^{2+}/Fe^{2+} mass ratio in the electreolyte [10].

2.2. Mechanism of electro deposition of Ni-Fe alloys

Electro deposition of Ni-Fe alloys exhibit the phenomenon of "anomalous codeposition". This term introduced by Brenner [11] is being used to describe the preferential deposition of the less noble metal, Fe, to the more noble metal, Ni. In other words, the reduction of nickel is inhibited while the deposition of iron is enhanced when compared with their individual deposition rates. According to Afshar et al. [12], the electrode position of nickel-iron alloys is a diffusion-controlled process with typical nucleation mechanism. According to Krause et al. [13], the anomalous behavior was assumed due to precipitation of iron hydroxide on surface electrode that inhibits the nickel reduction.

$$Ni^{\circ} / nFe^{\circ} + H^{+} \rightarrow 1/2\ Fe^{2+} + H_{atom} - Ni^{\circ}/\ (n\text{-}1/2)\ Fe \tag{1}$$

$$H_{atom} - Ni^{\circ}/(n\text{-}1/2)\ Fe^{\circ} + R\text{-}Cl \rightarrow Ni^{\circ}/\ (n\text{-}1/2)\ Fe^{\circ} + R + H\text{-}Cl^{-} \tag{2}$$

$$H_{atom} - Ni^{\circ}/(n\text{-}1/2)\ Fe^{\circ} + H + \rightarrow 1/2\ Fe^{2+} + Ni^{\circ}/(n\text{-}1)Fe^{\circ} + H_{2}\ (side\ reaction) \tag{3}$$

$$Ni^{\circ}/n\ Fe^{\circ} + 2H^{+} \rightarrow Fe^{2+} + H_{2} + Ni^{\circ}/\ (n\text{-}1)\ Fe^{\circ}\ (side\ reaction) \tag{4}$$

$$O_{2}(g) + 2H_{2}O + 4e \rightarrow 4OH^{-} \tag{5}$$

2.3. Phase formation

One of the important observations in the Ni-Fe alloys is the dependence of the crystal structure on the iron and nickel content in the deposited layers. Figure 3 shows X-ray diffraction patterns of various nanocrystalline Ni-Fe electrodeposits ranging in nickel content from 0 to 100%. Ni-Fe deposits with low nickel concentrations were found to have a body centered cubic (BCC) structure, while those with high nickel concentrations had a face-centered (FCC) structure [13]. While a mixed FCC/BCC structure was observed for nickel concentrations ranging from 10wt % to 40wt% nickel.

2.4. Properties of electrodeposited nanocrystalline Ni-Fe alloys

The first group of properties are strongly dependent on grain size. These include strength, ductility and hardness, wear resistance and coefficient of friction, electrical resistivity, coercivity, solid solubility, hydrogen solubility and diffusivity, resistance to localized corrosion and intergranular stress corrosion cracking, and thermal stability.

On the other hand, the second group of properties including bulk density, thermal expansion, Young's modulus resistance to salt spray environment, and saturation magnetization are little affected by grain size [5].

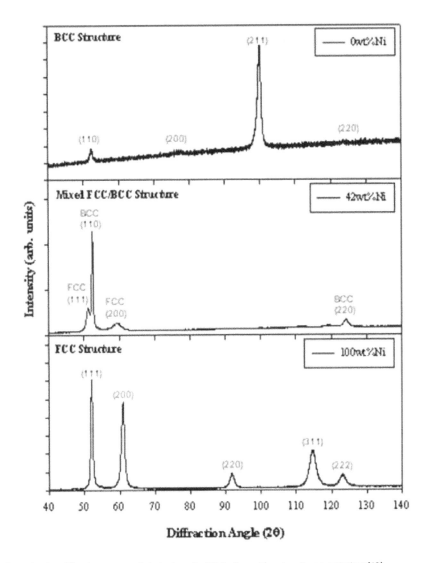

Figure 3. X-Ray diffraction patterns of electrodeposited Ni-Fe alloys with various Fe concentrations [13]

2.4.1. Mechanical properties

The plastic deformation behavior of electrodeposited nanocrystalline materials is strongly dependent on grain size. Initial increases, followed by significant decreases in hardness are observed with decreasing grain size (d) in the nanocrystal range, i.e., d ≤20 nm. The observed decreases in hardness are contrary to Hall-Petch behavior and consistent with results reported elsewhere for nanocrystalline materials. Others have only reported a reduction in the Hall-Petch slope in the nanometer range [5].

The grain size dependence of the proof stress was found to obey the Hall-Petch relationship; however, at constant grain size, lower values were always obtained with the equiaxed geometry, are shown in Table 2. In addition to the remarkable increases in hardness, yield strength, and ultimate tensile strength with decreasing grain size, it is interesting to note that the work hardening coefficient decreases with decreasing grain size to virtually zero at a grain size of 10 nm. The ductility of the material decreases with decreasing grain size from 50% elongation to failure in tension for conventional material to 15% at 100nm grain size and about 1% at 10 nm grain size. Generally somewhat greater ductility was observed in bending. A slight recovery in ductility was observed for grain sizes less than 10 nm. Compared to conventional polycrystalline Ni, nanocrystalline Ni electrodeposits exhibited drastically reduced wear rates and lower coefficients of friction as determined in dry air pin-on-disc tests. Contrary to earlier measurements on nanocrystalline materials prepared by consolidation of precursor powder particles, nanocrystalline nickel electrodeposits do not show a significant reduction in Young's modulus [4, 14]. This result provides further support for earlier findings of Krstic et al. [15], and Zugic et al. [16], which demonstrated that the previously reported reductions in modulus with nanoprocessing were likely the result of high residual porosity.

Property	Conventional	Nano Ni, 100 nm	Nano Ni, 10 nm
Yield strength, MPa (25 °C)	103	690	900
Yield strength, MPa (350 °C)	--	620	--
Ultimate tensile strength, MPa (25 °C)	403	1100	2000
Ultimate tensile strength, MPa (350 C)	--	760	--
Tensile elongation, % (25 °C)	50	15	1
Elongation in bending, % (25 °C)	--	40	--
Modulus of elasticity, GPa (25 °C)	207	214	204
Vickers Hardness, Kg/mm2	140	300	650
Work hadnening coefficient	0.4	0.15	0.0
Fatigue strength, MPa (108 cycles/air/ 25 °C)	241	275	--
Wear rate (dry air pin on disc, μm3/μm	1330	--	79
Coefficient of friction (dry air pin on disc)	0.9	--	0.5

Table 2. Mechanical properties of conventional and nanocrystalline Nickel

Due to Hall-Petch strengthening, nanocrystalline alloys offer significantly increased strength and hardness over conventional alloys. The table 3 summarizes tensile test data for Prem alloy (Fe-80% Ni- 4.8% Mo) and a nanocrystalline Ni-Fe alloy close to the Prem alloy in composition with average grain size between 10-15 nm. It is obvious that the yield strength, ultimate tensile strength and Vickers hardness values of the nanocrystalline Ni~20% Fe alloy significantly exceed those for the conventional Prem alloy. While the ductility represented in the elongation percentage of the conventional Prem alloy is much greater than that of the nanocrystalline

Ni~20% Fe alloy. Also figure 4 shows the Vickers hardness of nanocrystalline Ni-Fe alloys as a function of Iron content in the deposits along with the hardness values of for various conventional Ni-Fe alloys. The average hardness of the nanocrystalline Ni-Fe alloys is approximately 4 to 7 times higher than that of the conventional alloys as seen in table 3 and Figure 4. Besides, Figure 4 shows that there is a moderate decrease in the hardness with increasing the Iron content in the FCC range and a significant increase with Fe- content in the BCC range and a minimum hardness at the FCC-BCC transition [14].

Material	Yield strength 0.2% Offset MPa	Ultimate tensile strength	% Elongation	Vickers Hardness (VHN)
Nano-Ni-20% Fe	1785	2250	3-4%	550-600
Conv Premalloy	207	550	30%	100

Table 3. Mechanical properties of nanocrystalline Ni-20%Fe and conventional Prem alloy

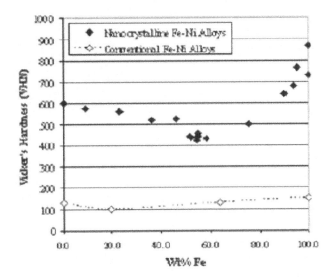

Figure 4. Vickers hardness as a function of iron content for various conventional and nanocrystalline Ni-Fe alloys [14].

2.4.2. Corrosion properties

In general, the corrosion resistance of nanocrystalline materials in aqueous solutions is of great importance in assessing a wide range of applications. To date, research in this area is still scarce and relatively few studies have addressed this issue. For the case of the corrosion behavior of nanocrystalline materials produced by crystallization of amorphous precursor materials; both beneficial and detrimental effects of the nanostructure formation on the corrosion performance were observed. The conflicting results are, to a large extent, due to the poorly characterized microstructures of the crystallized amorphous materials. On the other hand, for nanostructured materials produced by electro deposition, considerable advances in the understanding of microstructure on the corrosion properties have been made in recent years [18].

In previous studies, potentiodynamic and potentiostatic polarizations in de-aerated 2N H_2SO_4 (pH = 0) were conducted on bulk (2 cm^2 coupons, 0.2 mm thick) nanocrystalline pure Ni at grain sizes of 32, 50, and 500 nanometers and compared with polycrystalline pure Ni (grain size of 100 μm). Figure 5 shows the potentiodynamic anodic polarization curves of these specimens. The nanocrystalline specimens exhibit the same active-passive-transpassive behavior typical of conventional Ni. However; differences are evident in the passive current density and the open circuit potential. The nanocrystalline specimens show a higher current density in the passive region resulting in higher corrosion rates. These higher current densities were attributed to the higher grain boundary and triple junction content in the nanocrystalline specimens, which provide sites for electrochemical activity. However, this difference in current density diminishes at higher potentials (1100 mV SCE) at which the overall dissolution rate overcomes the structure-controlled dissolution rate observed at lower potentials [19].

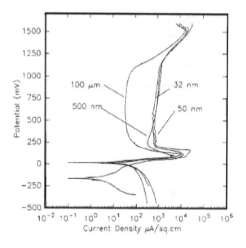

Figure 5. Potentiodynamic polarization curves for nanocrystalline and polycrystalline Ni in 2N H_2SO_4 at ambient temperature [19].

Figure 6 shows scanning electron micrographs of nickel with a) 32 nm and b) 100 μm grain size, held potentiostatically at 1200 mV (SCE) in $2NH_2SO_4$ for 2000 seconds. Both specimens exhibit extensive corrosion but the nanocrystalline Ni is more uniformly corroded while the specimen with 100 μm grain size shows extensive localized attack along the grain boundaries and triple junctions. X-ray photoelectron spectroscopy of the specimens polarized in the

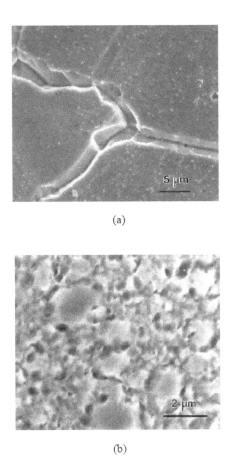

(a)

(b)

Figure 6. SEM micrographs of Ni with (a) 100 μm and (b) 32 nm grain size held potentiostatically at 1200 mV (SCE) in 2N H₂SO₄ for 2000 seconds [19].

passive region proved that the passive film formed on the nanostructured specimen is more defective than that formed on the polycrystalline specimen, while the thickness of the passive layer was the same on both specimens. This higher defective film on the nanocrystalline specimen allows for a more uniform breakdown of the passive film, which in turn leads to a more uniform corrosion. In contrast, in coarse-grained Ni the breakdown of the passive film occurs first at the grain boundaries and triple junctions rather than the crystal surface, leading to preferential attack at these defects [19].

The corrosion behavior of nanocrystalline Ni was also studied in 30 wt% KOH solution and pH neutral solution containing 3 wt% sodium chloride. The results were similar to the corrosion behavior observed in sulfuric acid. The general corrosion was somewhat enhanced compared to conventional polycrystalline Ni; however, the nanostructured materials were much more immune to localized attack which often can lead to catastrophic failures.

Comparing bulk nickel to nanocrystalline nickel, it is found that the bulk nickel was more resistant to anodic dissolution once the free corrosion potential had established. This interesting result because it indicates that once free corrosion conditions have been established, the surface of nanocrystalline nickel is more susceptible to corrosion than bulk nickel. It is known that the primary passivation potential of binary Ni-Fe alloys generally increase with increasing nickel concentration, comparatively little study has been conducted on the corrosion behavior of these alloys in nanocrystalline form. A study on the pitting behavior of nanocrystalline Ni-18% Fe found that it was more susceptible to pitting corrosion after significant grain growth had occurred during annealing [20]. Another study of the corrosion resistance of electrodeposited nanocrystalline Ni-W and Ni-Fe-W alloys reported poor corrosion resistance for the ternary alloy because of preferential dissolution of Fe. While alloy concentration effects on the corrosion rate of electrodeposited nanocrystalline Ni-Fe alloys remain to be clearly established, as the Iron content in the alloy is increased, the corrosion rate is increased simultaneously [21].

2.4.3. Electric and magnetic properties

As the average grain size in the nanocrystalline materials is reduced to the extent that the domain wall thickness is comparable to the grain size, the coercively is found to dramatically decrease while for the permeability of such alloys will increase. Another consequence of the ultra-fine grain size of nanocrystalline materials is an increase in the electrical sensitivity over the polycrystalline materials due to the high volume of grain boundaries. The electrical resistivity of electrodeposited nanocrystalline Ni-Fe alloys has been found to increase considerably as the grain size decreases to less than 100 nm [5, 22, 23].

2.4.4. Coefficient of thermal expansion

The grain size reduction to about 10nm in fully dense electrodeposited material has no major effect on the thermal expansion. Comparing between the coefficients of thermal expansion of nanocrystalline Fe-43wt% Ni to that of Ni-Fe conventional alloys it was found that both have similar values [18].

2.5. Applications of Ni-Fe alloys

Nickel-iron alloys are of great commercial interest as a result of their low thermal expansion and soft magnetic properties.

2.5.1. Low thermal expansion applications of Ni-Fe alloys

The nanocrystalline Ni-Fe alloys are used in the integrated circuit packaging and shadow masks for cathode ray tubes where they require a low coefficient of thermal expansion and also additional strength would be beneficial. For the integrated circuit packaging materials, they require materials with thermal coefficient matched to those of silicon to prevent the formation of cracks, de-lamination and/or de-bonding of the different materials during thermal cycles to which the components are exposed [14].

In the color cathode ray tube televisions and computer monitors, the shadow mask is a perforated metal sheet which the electrons from the electron gun must pass through before reaching the phosphor screen. Its role is to ensure that the electron beam hits only the correct colored dots and does not illuminate more than the one that was intended. Only 20% of the electrons pass through the shadow mask, and the other 80% are being absorbed by the mask which leads to an increase in the temperature of the mask. The resulting thermal expansion can disturb the alignment between the apertures and the phosphor triads, leading to a distorted image. This effect is known as "doming [14].

The main advantages of using electrodeposited nanocrystalline over conventional Ni-Fe alloys -for the use in integrated circuit packaging or shadow masks- for cathode ray tubes include:

• Single step process to produce foils ranging in thickness from 150μm to 500μm;

• High mechanical strength and hardness (> 450VHN);

• Isotropic properties due to fine equiaxed grain structure;

• Improved chemical machining performance;

• Finer pitch possible due to a decrease in grain size and higher strength;

• High etch rates due to increased grain boundary volume fraction [14].

2.5.2. Magnetic applications of Ni-Fe alloys

For soft magnetic applications, such as electromagnetic shielding, transformers materials, read-write heads, and high efficiency motors, magnetic materials- that exhibit small hysteresis losses per cycle- are required. More specifically, materials that have high permeability, low coercivity, high saturation and remnant magnetization, high electrical resistivity (to minimize losses due to eddy current formation [23-25]

3. Nanocrystalline Nickel-Molybdnum alloys

Nanostructured Ni-Mo alloys functionalized with acrylic acid (AA) and dispersed in paint have shown a satisfactory performance as corrosion inhibitors which led them to be used as anticorrosive paints for petroleum industry applications [26]. Recently, few researches have been conducted to investigate the effectiveness of nanocrystalline Ni-Mo alloys as catalysts for steam reforming of hydrocarbons, such as T. Huang et al.[27] Ni-Mo bimetallic catalyst of 20.5 nm crystallite size showed high activity, superior stability and the lowest carbon deposition rate (0.00073 gc.gcat^{-1}.h^{-1}) in 600-h time on steam.

Moreover, owing to the nanocrystalline state of Ni-Mo alloys, a large number of active sites are provided which make these alloys perfect candidates as catalysts for hydrogen production from water electrolysis [28]. Nanocrystalline Ni-Mo alloys have been processed by different techniques such as RF magnetron sputtering, mechanical alloying and electrodeposition techniques. H. Jin-Zhao et al. [29] reported that nanocrystalline Ni-Mo alloys prepared by RF magnetron sputtering technique to be promising electrodes for hydrogen evolution reactions. While P. Kedzierzawski et al. [30] were able to produce successfully nanocrystalline Ni-Mo alloys by mechanical alloying method, as well as indicating the positive contribution of the large surface area in increasing the catalytic effect as electrodes for hydrogen production and decrease of the exchange current density.

Furthermore, several attempts have been made in order to produce nanostructured Ni-Mo alloys by electro deposition technique to be used as cathodes for hydrogen production from water electrolysis. This is because electro deposition is considered to be cheaper than other production techniques -being mentioned previously- from the aspect of initial capital investment and running costs [31, 32]. While it also require minor modifications to existing conventional plating lines to be able to produce nanocrystalline films or stand free objects, in addition to that scaling up is relatively easy and high production rates can also be achieved [33].

Ni-Mo deposits have been well known for their use as cathodes for hydrogen production from water by electrolysis as well as catalysts for hydrogen production by steam reforming of hydrocarbons [7]. Arul Raj and Venkatesan [34] showed an increased electrocatalytic effect of Ni-Mo electrodeposited alloys for the hydrogen evolution reaction than that showed by nickel and other nickel-based binary alloys such as Ni-Co, Ni-W, Ni-Fe, and Ni-Cr. In addition, Ni-Mo alloys are considered as highly corrosion resistant due to the good corrosion protection characteristics of molybdenum in non oxidizing solutions of hydrochloric, phosphoric, and hydrofluoric acid at most concentrations and temperatures and in boiling sulfuric acid up to about 60% concentration [35]. The nickel-molybdenum alloys normally containing 26–35wt% Mo are among the few metallic materials that are resistant to corrosion by hydrochloric acid at all concentrations and temperatures [36]. Electro deposition is one of the most promising techniques for producing nanostructure materials owing to its relative low cost compared to the other methods. Electro deposition produces nanocrystalline materials when the deposition parameters (e.g., plating bath composition, pH, temperature, current density, etc.) are optimized such that electrocrystallization results in massive nucleation and reduced grain growth [37, 38]. Due to better anticorrosive in several aggressive environments, mechanical and thermal stability characteristics of Ni-Mo alloys, the electro deposition of these alloys plays an important role. It is an example of the induced codeposition mechanism [39].

According to J. Halim et al [40], Ni-Mo nanocrystalline deposits (7–43 nm) with a nodular morphology (Figure 7) were prepared by electro deposition using direct current from citrate-ammonia solutions. They exhibited a single Ni-Mo solid solution phase. Increasing the applied current density led to a decrease of the molybdenum content in the deposited alloys, increase in crystallite size, and increase of the surface roughness. The highest microhardness value (285 Hv) corresponded to nanodeposits with 23% Mo. The highest corrosion resistance accompanied by relatively high hardness was detected for electrodeposits containing 15% Mo. Mo content values between 11 and 15% are recommended for obtaining better electrocatalytic activity for Hydrogen evolution reaction "HER" with the lowest cathodic Tafel's constant.

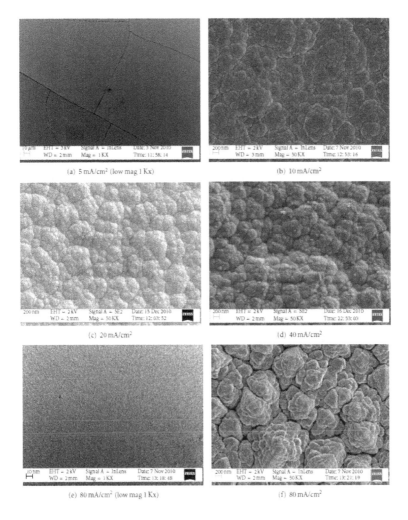

Figure 7. SEM micrograph of Ni-Mo nanocrystalline electrodeposits [40].

3.1. Mechanism of Ni-Mo electro deposition

Ni-Mo electrodeposited alloys are considered as an example of the induced codeposition mechanism. Several hypotheses have been proposed and many investigations have been carried out to describe the Ni-Mo electro deposition mechanism. According to Chassaing et al. [39], the formation of Ni-Mo electrodeposited alloy has been explained in the following steps:

1.   Reduction of molybdate to Mo (IV) oxide by Ni (II) species forming MoO_2Ni_4.

2.   Reduction of the mixed oxide (MoO_2Ni_4) forming Ni-Mo alloy.

Whereas, Podlaha and Landolt [41, 42], have proposed a model based on the adsorption and catalytic reduction of molybdate species using a bath containing citrate species. This model assumes that [Ni (II)HCit]⁻ catalyze molybdate reduction forming a surface-adsorbed intermediate, $[Ni(II)HCit-MoO_2]_{ads}$. The deposition of Mo, as indicated by this model, is only possible when Ni (II) ions are present, while nickel ions reduction following an independent path.

Generally, in a nickel-rich citrate electrolyte molybdenum deposition is mass-transport limited. Therefore, the Ni-Mo alloy composition is strongly influenced by the electrode rotation rate. While in a molybdenum-rich electrolyte, the rate of molybdenum deposition is limited by the flux of nickel ion. The alloy composition in this case is independent of hydrodynamic effects [41].

Electro deposition parameters have a large influence on the compostion of Ni-Mo alloys, the plating current density is one of the important deposition parameters affecting the composition of Ni-Mo alloys, and it has been investigated by many researches through different plating bathes that the molybdenum content decreases as the plating current density increases E. Chassaing et al. [39], reported the decrease of molybdenum content from 30 to about 10 wt% by increasing the current density from 10 to 150 mA/cm² as shown in the Figure 8 [42]. The same conclusion has been reached in the case of Ni-Mo alloys deposited from pyrophosphate baths as indicated by M. Donten et al. [31] that Mo content ranges from 55 to 35 wt% for a plating current density from 15 to 50 mA/cm².

As for the relationship between the Mo content in the citrate bath and that in the deposit, the molybdenum content in the deposit increases as the metal percentage of molybdenum in the bath increases [40]. The increase in the content of sodium citrate in the bath resulted in an increase of the molybdenum content of the deposit. Moreover, the molybdenum content in the deposit showed an increased as the pH increases till about pH = 9 and then decreased.

3.2. Phase formation of nanocrystalline Ni-Mo alloys

Most of the electrodeposited Ni-Mo alloys are composed of a single phase and are mainly semi amorphous. It was found by XRD analysis that the Ni-Mo alloys electrodeposited from a citrate bath (ph 8.5-9.5) contain Ni-Mo solid solution.

With diffraction peaks being sharp at the lower content of Mo (12wt%) and wide at the high concentration of Mo (30wt%) indicating that the alloys are tending to be more amorphous as the molybdenum content increases as shown in Figure 9 [42]. A similar conclusion was made by the XRD analysis of the Ni-Mo alloys electrodeposited from pyrophosphate-ammonium-chloride bath (pH 8.5) [31].

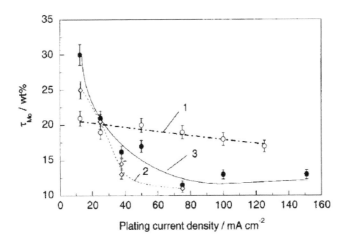

Figure 8. Molybdenum content as a function of the applied current density. Curve 1: 0.02 M Na₂MoO₄, pH 8.5; Curve 2: 0.02 M Na₂MoO₄, pH 9.5; Curve 3: 0.03M Na₂MoO₄ pH 9.5 [42].

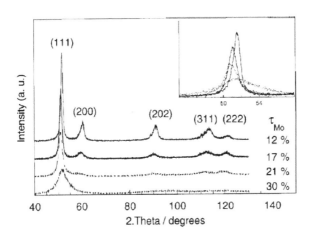

Figure 9. XRD patterns for Ni–Mo layers deposited from solution containing 0.03 M Na₂MoO₄ at pH 9.5 [42].

3.3. Properties of nanocrystalline Ni-Mo alloys

3.3.1. Corrosion properties of nanocrystalline Ni-Mo alloys

Corrosion resistance of nanocrystalline materials is of a great importance in assessing a wide range of potential future applications. To date, a few researches have addressed this issue. Ni-Mo alloys have been studied for their corrosion resistance especially in the media where they are used as cathodes for hydrogen evolution [34]. As indicated previously, Ni-Mo alloys are of excellent corrosion resistance especially in the aqueous solutions containing chloride ions [31]. As for Ni-Mo nanocrystalline alloys, they showed promising protective characteristics evidenced by polarization resistance of about 3.5-17 kΩ. While polarization curves being obtained for Ni-Mo nanocrystalline alloys being immersed in 0.5 M NaCl for various periods as shown in Figure 10 and 11, indicated that at the beginning the alloy with higher Mo content (27-30wt%) having a less negative potential (-0.014V) vs. Ag/AgCl than that for the alloy containing lower Mo content (13-15 wt%) recording a potential of about -0.749V vs. Ag/AgCl. As the immersion period increased, the corrosion potential moved towards more electronegative values, down to about -0.93 V vs. Ag/AgCl, regardless the alloy composition [43]. While in the deaerated hydrochloric solutions, nanocrystalline Ni-Mo alloys being deposited from citrate solution (pH 8.5-9.5) show a large passivation domain without any pitting. The corrosion currents as well as the passivation currents were higher than for the bulk conventional Ni-Mo alloy (Hastelloy B) and decreased when the Mo content in these alloys is increased [43].

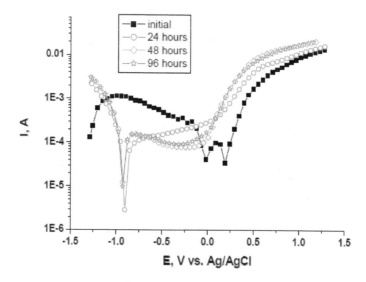

Figure 10. Polarization curves in semilogarithmic coordinates for Ni-Mo alloy (27-30% Mo) deposit in 0.5 M NaCl for various periods of continuous immersion (25 °C, 5 mV/s; geometrical surface of WE-0.63585 cm²) [43].

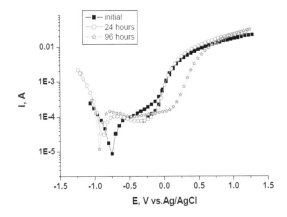

Figure 11. Polarization curves in semilogarithmic coordinates for Ni-Mo alloy (13-15% Mo) deposit in 0.5 M NaCl for various periods of continuous immersion (25 °C, 5 mV/s; geometrical surface of WE-0.63585 cm²) [43].

3.3.2. Electrocatalytic reactivity for hydrogen evolution

Nanocrystalline Ni-Mo alloys are known for their premium electrocatalytic properties as the presence of molybdenum in these materials apparently increases their catalytic activity. Ni-Mo alloy coatings electrodeosited from pyrophosphate-sodium bicarbonate bath possess high catalytic activity for hydrogen evolution in the NaOH solutions. Their stability in the 1 M NaOH at 25°C under the condition of the reverse polarization are of much better activity toward hydrogen evolution reaction than pure Ni electrode as the overpotential for Ni-Mo electrodeposited alloys was recorded as -1.3V vs. SCE at j = -0.3 Acm⁻² compared to -1.66 V vs.SCE for Ni electrode [44, 45].

4. Electrodeposition of Ni-Mo nano composites

Broad application of nickel-based composite coatings in electrochemistry is due to the highly catalytic activity in electrocatalytic hydrogen evolution (HER) and electrocatalytic oxygen evolution (OER) as well as good corrosion resistance of nickel in aggressive environments [46]. In order to improve the properties of these materials, and enhance their catalytic activity, various modifications could be applied, such as alloying with other elements, incorporating composite components. All these modifications aimed at obtaining the electrodes with very developed, rough or porous electrode surface. Many types of particles were used to improve mechanical, physicochemical or electrocatalytic properties of composite coatings, like carbides, silicides, nitrides, and oxides [47, 48]. Recent research in electro deposition suggested a technique involving codeposition of metallic particles to form electrodeposited metal matrix/ metal particle composites [49]. Among these composites are those containing metals like Al, Ti, V, Mo which could not be directly deposited from aqueous solutions. Incorporating such powdered components to the metal matrix leads to obtain a new kind of composite material which could be applied as electrode materials [50- 53]. Ni-based binary composite coatings like Ni–Mo, Ni–Zn, Ni–Co, Ni–W, Ni–Fe and Ni–Cr were tried for hydrogen electrodes; out of these electrodes, Ni–Mo was found to be best and most stable electrode with an overpotential of 0.18 V in 6 M KOH solutions [54]. Also according to Kubisztal et al. [55] electrolytic Ni–Mo and Ni–Mo–Si coatings were prepared by codeposition of nickel with silicon and molybdenum powders from a nickel bath in which Mo and Si particles were suspended by stirring. Composite coatings are characterized by very porous surface in comparison with nickel coating after the same thermal treatment. Ni–Mo composite coatings, obtained by electrodeposition of Ni with Mo particles on a steel substrate from the nickel bath containing suspended Mo powder, showed pronounced improvement in the electrochemical performance for HER in an alkaline environment compared to nickel electrode [47, 55]. The molybdenum content and the thickness of the Ni–Mo composite coatings change between 28–46 wt%, and 100–130 nm, respectively depending on the deposition current density. Ni–Mo electrodeposits were characterized by larger surface than the Ni electrodeposits. According to Jovic et al. [56], improved performances are to be expected if a composite compact layer of Ni and some Mo oxides could be prepared by the simultaneous electrodeposition of Ni and MoO_3 from an electrolyte solution in which MoO_3 particles are suspended. It seems a unique way for solving the problem of porosity of electroplated Ni–Mo coatings and low mechanical stability of thermally prepared Ni–Mo catalysts.

Ni–Mo nanocomposite coatings (18–32 nm) were prepared by electrodeposition of nickel from a nickel salt bath containing suspended Mo nanoparticles (Figure 12), by Abdel-Karim et al [57]. All the coatings have been deposited under galvanostatic conditions using current densities in the range 5–80 mA/cm^2. According to structural investigation carried out by X-ray diffraction, the obtained coatings consisted of crystalline Mo phase incorporated into Ni matrix (Figure13). The molybdenum content diminished with increasing the deposition current density and ranged between ~6 and ~17% Mo. Table 4 summarizes the corrosion properties of Ni-Mo composites examined in a remarkable deterioration in the corrosion resistance of Ni–Mo composites was observed with the increase of Mo content due to crystallite size-refining and surface roughness effect. Electrocatalytic effect for hydrogen production was improved mainly as a result of increasing the surface roughness and thus providing more accessible surface area.

According to Low and Walsh [58], many operating parameters influence the quantity of incorporated particles, including current density, bath agitation and electrolyte composition. It has been suggested that the transport of particles is due to electrophoresis, mechanical entrapment, adsorption, and convection–diffusion. The current density has been found to influence the amount of alumina nanoparticles incorporated into electrode-posited nickel. An increase in current density resulted in a rough surface microstructure and lead to less Al_2O_3 nanoparticles being incorporated in the metal deposit. The trend of decreasing the molybdenum content in Ni–Mo composites as a result of raising the applied current density has been also reported by Kubisztal et al. [55]. This phenomenon could be due to the fact that the amount of Mo powder incorporated in the composite depends on the rate of mass transport. At lower current densities Mo powder has longer time to reach the Ni matrix and form Ni–Mo composite. Whereas, at higher current densities, the Ni layers grow faster and less time is allowed for Mo powder to be transferred to the Ni matrix [54]. Another explanation for the same phenomenon is that by increasing the current density, a strong electric field is produced which in turn creates a partial desorption of nickel ions on the surface of Mo particles [59, 60]. The Ni crystallite size was reduced from ~32 to ~18 nm as the current density was decreased from 80 to 5 mA/cm^2. These values are less than that for Ni crystallites (62 nm) reported by Panek and Budniok [59]. This fact could be explained by the incorporation of metallic powder particles into the nickel matrix and the influence of the deposition current density. When electrically conductive parti-cles are embedded into the coating, they effectively become a part of the cathode and act as nucleation sites for the Ni matrix. A structure refinement occurs relative to pure Ni coatings due to the periodic nucleation and growth from the surface of the particles.

The generally accepted mechanism of the HER in alkaline solution is based on the following steps:

$$M + H_2O + e \rightarrow MH_{ads} + OH^- \tag{6}$$

$$MH_{ads} + H_2O + e \rightarrow H2 + M + OH^- \tag{7}$$

$$2\,MH \rightarrow H_2 + 2\,M \tag{8}$$

where (1) is the proton discharge electrosorption (Volmer reaction), which is followed by electrodesorption step (2) (Heyrovsky reaction) and/or H recombination step (3) (Tafel reaction). According to general models for HER mechanisms, the Volmer–Heyrovsky and Tafel reaction mechanism, reactions (1)–(3), may display three different slopes: 116.3, 38.8 or 29.1 mV dec^{-1} at 20 °C. When Heyrovsky reaction is the rate determining step, r.d.s., the slope is 38.8 mV dec^{-1} and when the Tafel reaction is the r.d.s. slope is 29.1 mV dec^{-1}. When the slope of 116.3 mV dec^{-1} is observed, it is impossible to distinguish which step is the rate determining one. However, as the surface coverage by adsorbed hydrogen should increase with the increase in negative overpotential, the rate limiting step should be that of Heyrovsky [54, 60].

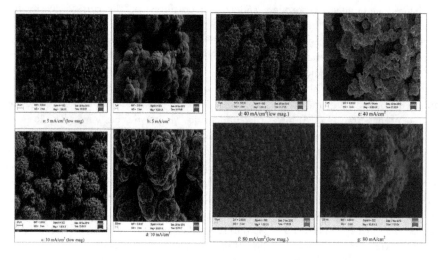

Figure 12. SEM micrographs for electrodeposited layers of Ni–Mo composites [57].

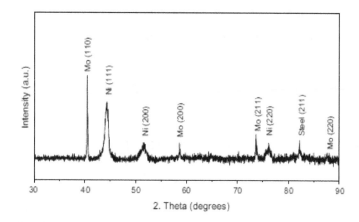

Figure 13. XRD of electrodeposited Ni-Mo nanocomposites [57].

4.1. Corrosion properties of electrodeposited Ni–Mo composites

From Table 4, electrodeposited Ni–Mo composites prepared by R. Abdel-karim et al et al [57] showed higher corrosion rates (0.035–1.795 mm/y) compared to the corrosion rates of Ni–Mo electrodeposited alloys (4.7–8.3 ×10^{-3} mm/y) [9]. This result would be attributed to the non-homogeneity and the rough surface of the Ni–Mo composites compared to the surface of Ni–Mo alloys. The highest electrocatalytic activity is detected for Ni–Mo composite containing

~6% Mo as well as the high surface roughness value (5.35 μm). The obtained results are in agreement with the results obtained by Panek and Budniok [59] and would indicate that both factors (Mo content and surface roughness) influence the electrocatlaytic behavior of Ni–Mo composites. The decrease of one of these two factors would lead to reduction in the electrocatalytic activity as for hydrogen evolution reaction. According to Kubisztal et al. [55], the main contribution toward the electrocatalytic activity of Ni–Mo composites is attributed to the increase in the actual surface area due to higher surface roughness. Mo particles, while embedded in the Ni matrix, produce a moderate increase in the catalytic effect in the point of contact.

Compared to previous work [40] and from table 5 [57], the Ni–Mo composite offers a higher electrochemical activity toward hydrogen evolution reaction (HER) than that of Ni–Mo alloys, with the sample containing 17% Mo had the lowest overpotential (−0.725 mV) at j = −1.5 mA/cm^2. These observations are in agreement with Popczyk [60].

Plating current density, mA/cm^2	Mo content, %wt	Ecorr, mV	Icorr, μm/cm^2	Corrosion rate, mm/y
5	14.6	-435	42.7	0.5
10	17.22	-512	153.5	1.795
20	16.7	-752	105.6	1.236
80	6.4	-356	2.94	0.035

Table 4. Corrosion properties of electrodeposited Ni-Mo nanocrystalline composites

Plating current density, mA/cm^2	Mo content, %wt	Roughness, Rq, μm	Ni crystallite size, nm	Cathodic mV/decade	Evs SCE(V) at j= -1.5 mA/cm^2
5	14.6	0.965	18.3	-115.2	-1
10	17.2	25.6	25.6	-142.7	-0.725
20	16.7	28.5	28.5	-118	-0.92
80	6.35	32.1	32.1	-34.9	n.a

Table 5. Electrocatalyric properties of electrodeposited Ni-Mo nanocrystalline composites

5. Nano-porous alumina formation

Nanoporous alumina films "AAO" produced during electrochemical anodization of aluminum has been studied for many years, and has continued to attract interest from various researchers because of its unique chemical and physical properties. These unique properties have made it possible for a wider application such as electronic devices, magnetic storage disks, sensors in hydrogen detection, adsorption of volatile organic compounds; biodevices; and in drug delivery, etc [61- 63].

The advantages of using the AAO membrane as a template are firstly, that it allows the diameter of the nanowires, nanorods and nanotube to be tailored to the respective pore size in the membrane (Figure 14) [63]. Secondly, it ensures that the growth of the nanocrystal (nanowire, nanorod, nanotube) is aligned within the high aspect ratio nano-channel which is also perpendicular to the substrate surface at the base of the membrane. A wide variety of materials that include metals, oxides, conductive polymers and semiconductors can then be deposited into the pores of the membrane. Then a suitable formation mechanism can be used to generate nanowires, nanorods (short nanowires) and nanotubes. The dimensions of which can be controlled by adjusting the template pore geometry and the formation parameters [64].

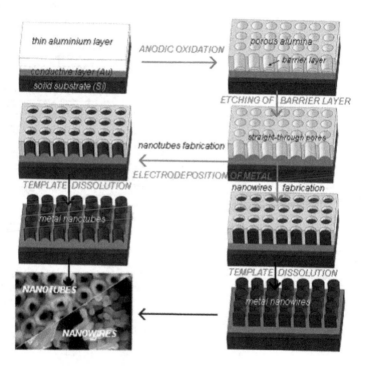

Figure 14. Template Assisted Growth of Nanowires and Nanorods [63]

Electrodeposition of nanoscale materials in the pores of a self- ordered nanochannel matrix is of relatively low cost. Furthermore, electrodeposition is a fast process for the fabrication of large arrays of nanostructures with a very large aspect ratio, which is not possible with standard lithographic techniques [65]. Electrochemical synthesis in a template is taken as one of the most efficient methods in controlling the growth of nanowires and has been used to produce a variety of metal nanowires [66].

AAO template is an ideal mold because it possesses many desirable characteristics, including tunable pore dimensions, good mechanical property, and good thermal stability. Moreover, especially for metal nanowires, the AAO template has been proved to be a cheap and high yield technique to produce large arrays of metals nanowires by the electrodeposition. This oxide has ordered hexagonal cells, of which every cell contains a cylindrical pore at its center. The pore diameter, cell size and barrier layer thickness can be controlled by anodizing voltage and the depth of pore by anodizing time [61]. Masuda et al. [64] has shown that AAO membranes with large pore sizes can be used in optical devices, particularly in the infra-red region of the spectrum.

The controlled placement of nano-dots on a substrate can be achieved with great accuracy using electron beam lithography. The major disadvantages of using this technique are the high capital cost of the equipment and the long exposure times which result in a low production output. An alternative process for the large scale fabrication of nano-dots on a semiconductor substrate uses an AAO membrane as an evaporation mask. The membrane provides an array of uniformly sized pores and site controlled pore locations where the nano-dots can be deposited. Before being used as a mask, the barrier layer is removed from the AAO membrane using a suitable etching technique, leaving a through pore structure or nano-channel. The major advantage of manufacturing nano-dot arrays using an AAO membrane is that the membrane can be used as a template. The template provides the precise location of individual pores in a high-density pore array without the need for expensive lithographic processes. The potential technological application of nano-dots is in the fabrication of ultra-small electronic devices, ultrahigh-density recording media, and nano-catalysis applications [64].

5.1. Nano-porous alumina formation mechanism

The formation process of the nano-porous oxide layer is a complex process that produces a self- organized hexagonal pore array, these hexagonal honeycomb structures have been reported by several researchers [67-73]. The porous structure consists of a thin non porous oxide layer of constant thickness that is adjacent to the metal substrate that continually regenerates at the base of the pore while the pore wall is being created, this wall increases in height with time. The particular electrolyte, its concentration, the anodic voltage and bath temperature are the main parameters in determining the pore size and the distance between pores [74].

Typical electrolytes used to produce this type of oxide layer have a pH that is less than 5, and slowly dissolve the forming oxide layer. Examples of the acids used are sulfuric, phosphoric and oxalic. However; mixtures of organic and inorganic acids have also been used. The properties of the electrolyte are important in the formation of porosity and permeability. Electrolytes that are composed of less concentrated acids tend to produces oxide coatings that are harder, thicker, less porous and more wear resistant than those composed of higher concentrated acids. But the most important factor that must be considered when forming the porous oxide is the electrolyte's ability to sustain a significant flow of Al^{3+} ions from the metal substrate into the electrolyte. There are two mechanisms that are responsible for the loss of Al^{3+} ions from the metal substrate. The first is by the direct expulsion of ions by the applied electrical field and the second is the dissolution of the forming oxide layer. In addition, if there are regions of high current flows when the electric field is applied, an increased dissolution rate can result (field assisted dissolution) [75].

The origin of pore nuclei and the exact mechanism of pore nucleation are still largely unknown. Several formation models have been proposed [76, 77]; one model explains that pore nucleation results from an electric field assisted local chemical dissolution [78] at the electrolyte/oxide interface and oxide generation at the metal/oxide interface. For electrolytes with a pH less than 5, there is a significant flow of Al^{3+} ions into the electrolyte and as a consequence there are regions where the formation of new oxide at the oxide/electrolyte interface is unstable. This regional instability produces variations in the applied electrical field; this in turn results in an increased dissolution rate [78, 79]. This mechanism produces an underlying metal/oxide and oxide/electrolyte interfaces that consist of a large numbers of hemispherical depressions per cm^2 that corresponds to the pores density. In these depressions the electrical field tends to be more concentrated due to the focusing effect of the hemispherical shape, hence the increased dissolution rate. [80]. In contrast, the electrical field is fairly constant over the surface of the barrier layer formed in an electrolyte with a pH greater than 5, where the oxide thickness is uniform and stable. These hemispherical depressions form the foundations of the resulting pore structures. The location of these depressions is also influenced by the initial surface topography, surface imperfections such as impurities, pits, scratches, grain boundaries and surface treatments prior to anodization [81]. Patermarakis et al. [82] proposed a pore nucleation model that results from the spontaneous recrystallization of the unstable rare lattice of oxide formed at the surface of the Al adjacent to metal/oxide interface to a more stable denser nano-crystalline oxide located in the oxide layer. The resulting recrystallization ruptures in the surface and produces regions of rarefied oxide between nanocrystallites (anhydrous/amorphous). It is in these regions that pore nuclei form. Earlier studies by Habazaki et al. [77] indicated the potential for the enrichment of alloying elements, dopants and/or impurities in the Al substrate adjacent to the metal/oxide interface. The enrichment layers were found to be about 1 to 5 nm thick immediately beneath the metal/oxide interface and were a consequence of the oxide growth. These enrichment layers may also be involved in the initiation of changes within the oxide layer that promote pore nucleation. In a recent investigation by Zaraska et al. [78], the presence of alloying elements in an Al alloy (AA1050) not only slowed the rate of oxide growth but also influenced structural features such as porosity, barrier layer thickness, pore diameter and pore density of the forming oxide layer.

In the early stages of the anodization process Al^{3+} ions migrate from the metal across the metal/oxide interface into the forming oxide layer [83]. Meanwhile O^{2-} ions formed from water at the oxide/electrolyte interface travel into the oxide layer. During this stage approximately 70% of the Al^{3+} ions and the O^{2-} ions contribute to the formation of the barrier oxide layer [79], the remaining Al^{3+} ions are dissolved into the electrolyte. This condition has been shown to be the prerequisite for porous oxide growth, in which the Al-O bonds in the oxide lattice break to release Al^{3+} ions [80]. During the oxide formation the barrier layer constantly regenerates with further oxide growth and transforms into a semi-spherical oxide layer of constant thickness that forms the pore bottom, as shown in Figure 15 [75].

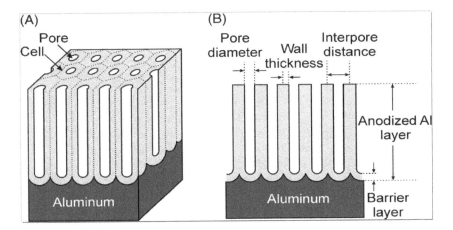

Figure 15. Schematic structure of anodic porous alumina (A) and a cross-sectional view of the anodized layer (B) [75].

During the formation of the porous oxide layer the anodic Al dissolution reaction is presented by:

$$2Al \rightarrow 2Al^{3+} + 6e- \tag{9}$$

$$6H+ + 6e- \rightarrow 3H_2 \tag{10}$$

$$2Al + 3O^{2-} \rightarrow Al_2O_3 + 6e- \tag{11}$$

$$2Al^{3+} + 3H_2O \rightarrow Al_2O_3 + 6H+ \tag{12}$$

Sum of the separate reactions at electrode: (Overall anodization of Al equation)

$$2Al + 3H_2O \rightarrow Al_2O_3 + 3H_2 \tag{13}$$

The steady state growth results from the balance between the field-enhanced oxide dissolution at the oxide/electrolyte interface at the base of the hemispherical shaped pores where the electric field is high enough to propel the Al^{3+} ions through the barrier layer and the oxide growth at the metal/oxide interface resulting from the migration of O^{2-} and OH^- ions into the pore base oxide layer, as in Figure 16. This also explains the dependence of the size of the pore diameter to the electric field produced by the anodizing voltage. It should also be noted that the electric field strength in the pore walls is too small to make any significant contribution to the flow of ions.

The oxidation takes place over the entire pore base and the resulting oxide material grows perpendicular to the surface, neighboring pore growth prevents growth in any other direction. The vertical growth of the pore wall creates a columnar structure with a high aspect ratio that contains a central circular channel. This channel extends from the base of the pore to the surface of the oxide layer [81]. This upward growth of the pore wall was recently investigated by Garcia-Vergara et al. [86] in which a tungsten tracer was placed into an initial oxide layer formed by an initial anodization step. During the next stage of anodization, the position of the tracer was monitored and found to travel from the metal/oxide interface of the barrier layer located at the base of the pore towards the growing wall structure. This flowing motion of the tracer was credited to the mechanical stresses being generated by the continued formation of new oxide within the pore base and the repulsive forces set up between neighboring pores during the growth of the wall structure. These forces resulted from the volume expansion (by a factor of 2) during the oxidation of Al to alumina. This volumetric expansion factor results from the difference in the density of Al in alumina (3.2 g cm^{-2}) and that of metallic Al (2.7 g cm^{-2}). This volumetric oxide expansion at the metal/oxide interface also contributes to the hemispherical shape of the pore base. However, under normal experimental conditions, the volumetric expansion factor is less than 2. This is due to the hydration reaction that occurs at the oxide/electrolyte interface which results in the dissolution and thinning of the oxide layer.

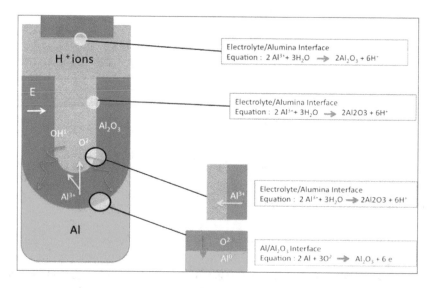

Figure 16. Schematic of ion movement during pore formation [81].

The volumetric growth of the oxide layer is also dependent upon the type of electrolyte (this effect is more evident when using phosphoric or sulfuric acids), the electrolyte concentration and the electric field created by the anodizing voltage. Furthermore, a research investigation by Li et al. [85] has shown that a volume expansion factor of 1.4 could be achieved under optimal anodizing parameters, independent of the electrolyte used. The porous oxide layer thickness and hence pore height can grow to many times the height of the barrier layer. It is usually during this oxide growth that electrolyte anions can integrate into the forming porous structure near the oxide/electrolyte interface, while pure alumina is predominantly found in layers close to the metal/oxide interface. It is also possible for voids to form in the pore walls during the growth of the oxide layer. Possible causes of these voids range from oxygen evolution during oxide formation to localized defects in the barrier layer. These defects produce a condensation effect that involves cations and/or metal vacancies at the metal/oxide interface, which subsequently become detached, and form the void.

Diggle et al. [86].discussed the contradictory research results of the period. For example, the non- porous barrier layer was regarded as amorphous and anhydrous, while the porous layer had been found to be both amorphous and crystalline. In the case of the barrier layer, under normal anodization conditions the layer will be amorphous. However, studies by Uchi et al. [87] have shown that with the right growth conditions it was possible to have amorphous or crystalline Al oxide being produced during anodization. To form a crystalline oxide structure; an Al substrate was first immersed in boiling water to form a hydrous oxide layer [oxy-hydroxide with excess water (AlOOH.H$_2$O)]. The substrate was then anodized in a neutral borate solution at high temperatures, during which Al^{3+} ions move from the metal substrate to the hydrous oxide interface, where they combine and transform the hydrous oxide to crystalline Al$_2$O$_3$. Some of the contradictory evidence of the earlier works discussed by Diggle et al. [86] of the porous layer could be explained by the work of De Azevedo et al. [88]. In this study the structural characteristics of doped and un-doped porous Al oxide, anodized in oxalic acid was investigated using X-ray diffraction (XRD). The XRD patterns for the un-doped samples revealed several peaks associated with Al and Al$_2$O$_3$ crystalline phases on top of a broad peak that was approximately centered on the 2θ angle of 25°. This broad peak indicated that the synthesized layer was a highly disordered and/or amorphous Al oxide compound.

6. Conclusions

- One of the most important aspects of self-assembly lies in the capability of producing uniform structures over a large area using conventional electrodepsotion and anodizing processes.

- Nanostructured Ni-Fe alloys, produced by electro-deposition technique provide material with significant improved strength and good magnetic properties, without compromising the coefficient of thermal expansion.

- Due to better anticorrosive in several aggressive environments, mechanical and thermal stability characteristics of Ni-Mo alloys, the electro deposition of these alloys plays an important role. Broad application of nickel-based composite coatings in electrochemistry is due to the highly catalytic activity in electrocatalytic hydrogen evolution (HER) and electrocatalytic oxygen evolution (OER) as well as good corrosion resistance of nickel in aggressive environments.

• AAO template has been proved to be a cheap and high yield technique to produce large arrays of metals nanowires with tunable pore dimensions, good mechanical property, and good thermal stability by the electrodeposition.

Author details

R. Abdel-Karim[1] and A. F. Waheed[2]

1 Department of Metallurgy, Faculty of Engineering, Cairo University, Giza, Egypt

2 Department of Metallurgy, Nuclear Research Center, Cairo, Anshas, Egypt

References

[1] Luther, W. Industrial Application of Nanomaterials- Chances and Risks, Future Technologies Division of VDI Technologiezentrum GmbH, (2004).

[2] Aliofkhazraei, M. Nanocoatings, chapter 1, Springer, (2011).

[3] Gleiter, H. Nanostructured Materials: Basic Concepts and Microstructure. Acta Materialia (2000). , 48(1), 1-29.

[4] Peuloro, S, & Lincot, D. Cathodic Electrodeposition from Aqueous Solution of Dense or Open-structured Zinc Oxide Films. Advanced. Materials (1996). , 8, 166-170.

[5] Koch, C. C. Nanostructured Materials: Processing, Properties and Applications, William Andrew Publishing; 2 nd edition, (2007).

[6] Bockris, J. O'M. Fundamental Aspects of Electrocrystallization, Plumn Press, New York, 3 rd edition, first published; (1967).

[7] Uve, E. Materials Processing Handbook, chapter 22, CRC; (2007).

[8] Seo, M. H, Kim, D. J, & Kim, J. S. The Effects of pH and Temperature on Ni-Fe-P Alloy Electrodeposition from a Sulfamate Bath and the Material Properties of the Deposits. Thin Solid Films (2005)., 489 (1-2), 122-129.

[9] Sam, S, Fortas, G, Guittoum, A, Gabouze, N, & Djebbar, S. Electrodeposition of NiFe Films on Si (1 0 0) Substrate, Surface Science (2007). , 601(18), 4270-4273.

[10] Abdel-karim, R, Reda, Y, Muhammed, M, Raghy, S, Shoeib, M, & Ahmed, H. Electrodeposition and Characterization of Nanocrystalline Ni-Fe Alloys. Journal of Nanomaterials 2011; (2011). Article ID 519274, 8 pages.

[11] Brenner, A. Electrodeposition of Alloys, chapter 1, Academic Press, New York, NY, USA; (1963).

[12] Afshar, A, Dolati, A. G, & Ghorbani, M. Electrochemical Characterization of the Ni-Fe Alloy Electrodeposition from Chloride-Citrate-Glycolic Acid Solutions", Materials Chemistry and Physics (2002)., 72(2), 352-258.

[13] Krause, T, Arulnayagam, L, & Pritzker, M. Model for Nickel-Iron Alloy Electrodeposition on a Rotating Disk Electrode. Journal of the Electrochemical Society(1997). , 144(6), 953-960.

[14] Mc Crea, J. L, Palumbo, G, Hibbard, G. D, & Erb, U. Properties and Applications for Electrodeposited Nanocrystalline Fe-Ni Alloys. Review of Advanced Material Science (2003). , 5, 252-258.

[15] Krstic, V, Erb, U, & Palumbo, G. Effect of Porosity on Young's Modulus of Nanocrystalline Materials. Scripta Metallurgia and Materiala (1993). , 29, 1501-1504.

[16] Zugic, R, Szpunar, B, Krstic, V. D, & Erb, U. Effect of Porosity on the Elastic Response of Brittle Materials: An Embedded-atom Method Approach. Philosophical Magazine A- Physics of Condensed Matter Structure Defects and Mechanical Properties (1997). , 75(4), 1041-1055.

[17] Brooks, I. Synthesis and Mechanical Properties of Bulk Quantities of Electrodeposited Nanocrystalline Materials, University of Toronto. Ph. D thesis, Canada, (2012).

[18] Sherik, M, Ajwad, H, Rasheed, A. H, Jabran, A, & Ortiguerra, R. G. Adhesion and Corrosion Performance of Nanostructured Nickel and Cobalt-based Coatings. COR-ROSION 2008, March New Orleans LA.; (2008). Document ID 08025, 16-20.

[19] Shangyu, W. Electrochemical Properties of Nanocrystalline Nickel and Nickel-Molybdenum Alloys. Ph.D dissertation, Queen's University, Kingston, Ontario, Canada; (1997).

[20] Alves, H, Ferreira, M. G. S, & Koster, U. Corrosion Behavior of Nanocrystalline (Ni70Mo30)90B10 Alloys in 0.8 M KOH Solution. Corrosion Science (2003). , 45, 1833-1845.

[21] Egberts, P, & Hibbard, G. D. Mesoscale. Compositionally Modulated Nanocrystalline Ni-Fe Electrodeposits for Nanopatterning Applications. Journal of Nanomaterials 2008; (2008). Article ID 858235.

[22] Ramanujan, R. V. Nanostructured Electronic and Magnetic Materials. Sādhanā (2003)., 1&2, 81-90.

[23] Suryanarayana, C, & Koch, C. C. Nanocrystalline Materials- Current Research and Future Directions. Hyperfine Interactions (2000). , 130, 5-44.

[24] Reda, Y. Preparation and Characterization of Electrodeposited Nano-crystalline Fe-Ni Alloys. Ph. D dissertation, Cairo University, Egypt; (2010).

[25] Halim, J. Electrodeposition and Characterization of Nanocrystaline Ni-Fe Alloys. B.Sc graduation project, Cairo University; Egypt (2009).

[26] Garibay, V, Diaz, L, Paniagua, A, & Palacios, E. Nanostructured Materials Development with Applications to the Petroleum Industry. Acta Microscopia (2000). , 18, 52-58.

[27] Huang, T, Huang, W, Huang, J, & Ji, P. Methane Reforming Reaction with Carbon Dioxide Over SBA-15 Supported Ni-Mo Bimetallic Catalysts, Fuel Process. Technology (2011). , 92(10), 1813-2126.

[28] Schulz, R, Huot, J, & Trudeau, M. Nanocrystalline Ni-Mo Alloys and Their Application in Electrocatalysis. Journal of Material Research (1994). , 9(11), 2998-3008.

[29] Jin-zhao, H, Zheag, X, & Hai-ling, L. Effects of Sputtering Conditions on Electrochemical Behavior and Physical Properties of Ni-Mo Alloy Electrode", Transactions of Nonferrous Metals Society of China (2006). , 16(5), 1092-1096.

[30] Kedzierzawski, P, Oleszak, D, & Janik-czachor, M. Hydrogen Evolution on Hot and Cold Consolidated Ni-Mo Alloys Produced by Mechanical Alloying. Materials Science and Engineering A (2001)., 300(1-2), 105-112.

[31] Donten, M, Cesiulis, H, & Stojek, Z. Electrodeposition of Amorphous/nanocrystalline and Polycrystalline Ni-Mo Alloys from Ayrophosphate Baths. Electrochemica Acta (2005). , 50(6), 1405-1412.

[32] Clark, D, Wood, D, & Erb, U. Industrial Applications of Electrodeposited Nanocrystals", Nanostructured Materials (1997)., 9(1-8), 755-758.

[33] Robertson, A, Erb, U, & Palumbo, G. Practical applications for Electrodeposited Nanocrystalline Materials. Nanostructured Materials (1999)., 12(5-8), 1035-1040.

[34] Raj, I. A, & Vekatesan, V. K. Characterization of nickel-molybdenum and nickel-molybdenum-iron alloy coatings as cathodes for alkaline water electrolysers. International Journal of. Hydrogen Energy (1988). , 13(4), 215-223.

[35] Palumbo, G, Gonzalez, F, & Brennenstuhl, A. M. In-situ Nuclear Steam Generator Repair Using Electrodeposited Nanocrystalline Nickel. Nanostructured Materials (1997)., 9(1-8), 737-746.

[36] Robertson, A, Erb, U, & Palumbo, G. Practical Applications for Electrodeposited Nanocrystalline Materials. Nano structured Materials (1999). , 12, 1035-1040.

[37] Sundaramurthy, V, Dalai, A. K, & Adjaye, J. Comparison of P-containing γ-Al2O3 Supported Ni-Mo Bimetallic Carbide, Nitride and Sulfide Catalysts for HDN and HDS of Gas Oils Derived from Athabasca Bitumen. Applied Catalysis A (2006)., 311(1-2), 155-163.

[38] Borowiecki, T, Gac, W, & Denis, A. Effects of Small MoO3 Additions on the Properties of Nickel Catalysts for the Steam Reforming of Hydrocarbons: III. Reduction of Ni-Mo/Al2O3 Catalysts. Applied Catalysis A (2004)., 270(1-2), 27-36.

[39] Chassaing, E, Roumegas, M. P, & Trichet, M. F. Electrodeposition of Ni-Mo Alloys with Pulse Reverse Potentials. Journal of Applied Electrochemistry (1995). , 25(7), 667-670.

[40] Halim, J, Abdel-karim, R, Raghy, S, Nabil, M, & Waheed, A. Electrodeposition and Characterization of Nanocrystalline Ni-Mo Catalysts for Hydrogen Production. Journal of Nanomaterials 2012; (2012). Article ID 845673, 10 pages.

[41] Podlaha, E. J, & Landolt, D. I. An Experimental Investigation of Ni-Mo Alloys. Journal of Electrochemical Society (1996). , 143(3), 885-892.

[42] Podlaha, E. J, & Landolt, D. II.A Mathematical Model Describing the Electrodeposition of Ni-Mo Alloys. Journal of Electrochemical Society (1996). , 143(3), 893-899.

[43] Chassaing, E, & Portail, N. Characterization of Electrodeposited Nanocrystalline Ni-Mo Alloys. Journal of Applied Electrochemistry (2005). , 34(1), 1085-1091.

[44] Prioteasa, P, & Anicai, L. Synthesis and Corrosion Characterization of Electrodeposited Ni-Mo alloys Obtained from Aqueous Solutions, U.P.B. Science Bulletin B (2010). , 72(4), 11-24.

[45] Krstajic, N. V, & Jovic, V. D. Electrodeposition of Ni-Mo Alloy Coatings and their Characterization as Cathodes for Hydrogen Evolution in Sodium Hydroxide Solution", International Journal of Hydrogen Energy (2008). , 33(6), 3676-3687.

[46] Niedbala, J. Surface Morphology and Corrosion Resistance of Electrodeposited Composite Coatings Containing Polyethylene or Polythiophene in Ni-Mo Base, Bulletin of Material Science (2011). , 34, 993-996.

[47] Panek, J, & Budniok, A. Study of Hydrogen Evolution Reaction on Nickel-Based Composite Coatings containing Molybdenum Powder", International Journal of Hydrogen Energy (2007). , 32, 1211-1218.

[48] Aruna, S. T. William Grips, V. K., Rajam K. S. Ni-based Electrodeposited Composite Coating Exhibiting Improved Microhardness, Corrosion and Wear Resistance Properties", Journal of Alloys and Compounds (2009). , 468(2009), 546-552.

[49] Lekka, M, Kouloumbi, N, Gajo, M, & Bonora, P. L. Corrosion and Wear Resistant Electrodeposited Composite Coatings. Electrochimica Acta (2005). , 50, 4551-4556.

[50] Naploszek-bilnik, I, Budniok, A, & Lagiewka, E. Electrolytic Production and Heat-treatment of Ni-based Composite Layers Containing Intermetallic phases. Journal of Alloys and Compounds (2004). , 382, 54-60.

[51] Ping, Z, Cheng, G, & He, Y. Ni-P- SiC Composite Coatings Electroplated on Carbon Steel Assisted by Mechanical Attrition. Acta Metallurgica. Sinica. (Engl. Lett.) (2010). , 23(1), 1-10.

[52] Raj, I. A, & Vasu, K. I. Transition Metal-based Cathodes for Hydrogen Evolution in Alkaline Solution: Electrolysis on Nickel-based Electrolytic Codeposites. Journal of Applied Electrochemistry (1992). , 22(5), 471-477.

[53] Jakšić JM., Vojnović M.V., Krstajić N.V. Kinetic Analysis of Hydrogen Evolution at Ni-Mo Alloy Electrodes. Electrochimica Acta (2000). , 45, 4151-4158.

[54] Raj, I. A. Nickel Based Composite Electrolytic Surface Coatings as Electrocatalysts for the Cathodes in the Energy Efficient Industrial Production of Hydrogen from Alkaline Water Electrolytic Cells. International Journal of Hydrogen Energy (1992). , 17(6), 413-421.

[55] Kubisztal, J, & Budniok, A. Study of the Oxygen Evolution Reaction on Nickel-based Composite Coatings in Alkaline Media. International Journal of. Hydrogen Energy (2008). , 33, 4488-4494.

[56] Jovic, B.M, Lacnjeva, U, Jovic, V, Serb, J, & Ni-MoO2 Composite Cathodes for Hydrogen Evolution in Alkaline Solution. Effect of Aging of the Electrolyte for Their Electrodeposition. Journal of Serbian Chemical Society (2012). , 77 (0): 1-21.

[57] Abdel-karim, R, Halim, J, Raghy, S, Nabil, M, & Waheed, A. Surface Morphology and Electrochemical Characterization of Electrodeposited Ni-Mo Nanocomposites as Cathodes for Hydrogen Evolution. Journal of Alloys and Compounds (2012). , 530, 85-90.

[58] Low, C. T, & Walsh, F. C. Electrodeposition of Composite Coatings Containing Nanoparticles in a Metal Deposit. Surface Coating Technology (2006). , 2006, 371-383.

[59] Panek, J, & Budniok, A. Electrochemical Production and Characterization of Ni-Based Composite Coatings Containing Mo Particles. Advanced Review of Material Science (2007). , 201(14), 6478-6483.

[60] Popczyk, M. The Hydrogen Evolution Reaction on Electroctrolytic Nickel-Based Coatings Containing Metallic Molybdenum. Materials Science Forum (2010). , 636-637. 1036-1041.

[61] Eddy, G, Poinern, J, Ali, N, & Fawcett, D. Progress in Nano-Engineered Anodic Aluminum Oxide Membrane Development. Materials (2011). , 4, 487-526.

[62] Hamrakulov, B. Kim In-Soo, Lee M. G., Park B. H. Electrodeposited Ni, Fe, Co and Cu Single and Multilayer Nanowire Arrays on Anodic Aluminum Oxide Template, Transaction of Nonferrous Metals Society of China (2009). , 19, 583-587.

[63] Vorozhtsova, M, Hrdy, R, & Hubalek, J. Vertically Aligned Nanostructures for Electrochemical Sensors. Rožnov pod Radhoštěm. NANOCON, 20 - 22. 10. (2009).

[64] Masuda, H, Yada, K, & Osaka, A. Self-ordering of Cell Configuration of Anodic Porous Alumina with Large-Size Pores in Phosphoric Acid Solution. Jpn. Journal of Applied Physics (1998). , 37, LL1340- 1342.

[65] Kim, J. R, Oh, H, So, H. M, Kim, J. J, Kim, J, Lee, C. J, & Lyu, S. C. Schottky Diodes Based on a Single GaN Nanowire. Nanotechnology (2002). , 2002(13), 701-704.

[66] Mikhaylova, M, Toprak, M, Kim, D. K, Zhang, Y, & Muhammed, M. Nanowire Formation by Electrodeposition in Modified Nanoporous Polycrystalline Anodic Alumina Templates. Mat. Research. Society. Symp. Proceeding (2002). , 704, 155-160.

[67] Cheng-min ShenXiao-gang Zhang, Hu-lin Li," DC Electrochemical Deposition of CdSe Nanorods Array Using Porous Anodic Aluminum Oxide Template", Materials Science and Engineering A (2001). , 303, 19-23.

[68] Li, A. P, Muller, F, Birner, A, Nielsch, K, & Gosele, U. Polycrytalline Nanopore arrays with Hexagonal Ordering on Aluminium. Journal of Vac. Science Technology. A (1999). , 17, 1428-1431.

[69] Li, A. P, Muller, F, Birner, A, Nielsch, K, & Gosele, U. Hexagonal Pore Arrays with a 50-420 Interpore Distance Formed by Self-organisation in Anodic Alumina. Journal of Applied Physics (1998). , 84, 6023-6026.

[70] Zhang, L, Cho, H. S, Li, F, Metzger, R. M, & Doyle, W. D. Cellular Growth of Highly Ordered Porous Anodic Films on Aluminium. Journal of Material Science Letter (1998). , 17, 291-294.

[71] Hou, K, Tu, J. P, & Zhang, X. B. Preparation of Porous Alumina Film on Aluminium Substrate by Anodization in Oxalic Acid. China Chemical Letters (2002). , 13, 689-692.

[72] Kim, Y. S, Pyun, S. I, Moon, S. M, & Kim, J. D. The Effects of Applied Potential and pH on the Electrochemical Dissolution of the Barrier Layer in Porous Anodic Oxide Film on Pure Aluminium. Corrosion Science (1996). , 38, 329-336.

[73] Sullivan, O, Wood, J. P, & Nucleation, G. C. and Growth of Porous Anodic Films on Aluminium. P Roy. Lond. A Mat. A (1970). , 317, 511-543.

[74] Thompson, G. E, Furneaux, R. C, Wood, G. C, Richardson, J. A, & Goode, J. S. Nucleation and Growth of Porous Anodic Films on Aluminium. Nature (1978). , 272, 433-435.

[75] Eftekhari, A. Nanostructured Materials in Electrochemistry. WILEY-VCH Verlag GmbH & Co. KGaA, Weinheim, chapter1, (2008).

[76] Hoar, T. P, & Yahaloom, J. The Initiation of Pores in Anodic Oxide Films Formed on Aluminium in Acid Solutions. Journal of Electrochemical Society (1963). , 110, 614-621.

[77] Habazaki, H, Shimizu, K, Skeldon, P, Thompson, G. E, Wood, G. C, & Zhou, X. Nanoscale Enrichments of Substrate Elements in the Growth of Thin Oxide Films. Corrosion Science (1997). , 39, 731-737.

[78] Zaraska, L, Sulka, G. D, Szeremeta, J, & Jaskula, M. Porous Anodic Alumina Formed by Anodisation of Aluminium Alloy (AA1050) and High Purity Aluminium. Electrochimica Acta (2010). , 55, 4377-4386.

[79] Palbroda, E. Aluminium Porous Growth-II on the Rate Determining Step. Electrochimica Acta (1995). , 40, 1051-1055.

[80] Shawaqfeh, A. T, & Baltus, R. E. Fabrication and Characterization of Single Layer and Muti-layer Anodic Alumina Membrane. Journal of Membrane Science (1999). , 157, 147-158.

[81] Essa, A. K. Porous Anodic Alumina Oxide (AAO) by One Step Anodization at low Voltage..Ph.D Thesis, Faculty of Engineering, Cairo University, Egypt (2013).

[82] Habazaki, H, Shimizu, K, Skeldon, P, Thompson, G. E, Wood, G. C, & Zhou, X. Nanoscale Enrichments of Substrate Elements in the Growth of Thin Oxide Films. Corrosion Science (1997). , 39, 731-737.

[83] Zaraska, L, Sulka, G. D, Szeremeta, J, & Jaskula, M. Porous Anodic Alumina Formed by Anodisation of Aluminium Alloy (AA1050) and High Purity Aluminium. Electrochimica Acta (2010). , 55, 4377-4386.

[84] Garcia-vergara, S. J, Iglesias-rubianes, L, Blanco-pinzon, C. E, Skeldon, P, Thompson, G. E, & Campestrini, P. Mechanical Instability and Pore Generation in Anodic Alumina. P. Roy Soc. Lond. A Mat. (2006). , 462, 2345-2358.

[85] Li, F, Zhang, L, & Metzger, R. M. On the Growth of Highly Ordered Pores in Anodized Aluminium Oxide. Chem. Mater. (1998). , 1998(10), 2470-2480.

[86] Diggle, J. W, Downie, T. C, & Goulding, C. W. Anodic Oxide Films on Aluminium. Chem. Review (1969). , 69, 365-405.

[87] Uchi, H, Kanno, T, & Alwitt, R. S. Structural Features of Crystalline Anodic Alumina Films. Journal of. Electron. Society (2001). 148, B17-B23.

[88] De Azevedo, W. M, De Carvalho, D. D, Khoury, H. J, & De Vasconcelos, E. A. Da Silva E.F. Jr. Spectroscopic. Characteristics of Doped Nanoporous Aluminium Oxide. Material Science Engineering B. (2004). , 112, 171-174.

Chapter 5

Plasma Electrolytic Oxidation Coatings on Lightweight Metals

Qingbiao Li, Jun Liang and Qing Wang

Additional information is available at the end of the chapter

1. Introduction

Lightweight metals, e.g. aluminum (Al), magnesium (Mg), titanium (Ti) and their alloys are of great importance for applications in various machinery and transportation system, especially in aerospace and automobile products due to their high strength-to-weight ratio and superior physical and chemical performances. However, their poor tribological properties, such as low wear resistance, high friction coefficient and difficulty to lubricate, have seriously restricted their extensive applications.

In the past decades, various traditional surface treatments, such as physical vapor deposition [1-2], chemical vapor deposition [3], ion beam assisted deposition [4] and spraying [5], have been applied to metallic substrates to improve their generally poor tribological properties. However, most of the aforementioned methods involve high processing temperature, which may degrade the coatings and/or substrates. Here, a relatively novel technique, plasma electrolytic oxidation (PEO) treatment used to improve the tribological properties of lightweight metals was introduced.

1.1. The origin of PEO technique

Plasma electrolytic oxidation (PEO), also called micro-arc oxidation (MAO) [6], micro-plasma oxidation (MPO) [7], anodic spark deposition (ASD) [8] or micro-arc discharge oxidation (MDO) [9] in modern scientific literatures, is derived from conventional anodizing [10-11]. Anodizing is traditionally carried out using direct current (DC) electrolysis. The workpiece is made anodic in an acid electrolyte (sulfuric acid is most commonly used, but phosphoric, oxalic, chromic and other acids can be used, singly or in combination). Typically, the cell voltage is 20 to 80 V DC and the current density is 1 to10 A dm^{-2}, the process usually being controlled at a constant cell voltage. Plasma electrolytic oxidation (PEO) treatment usually carried out in high voltage condition which is introduced into the high-pressure discharge area from the Faraday region of traditional anodizing. The applied voltage is increased from tens of volts to hundreds of volts, which is the breakthrough of traditional anodizing. The voltage forms developed from DC to continuous pulse, and then to AC, resulting in corona, glow, spark discharge and even micro-arc discharge phenomenon in the surface of the samples [12]. The general comparison between conventional DC anodizing and PEO technique was shown in Table 1.

Properties	Anodizing	PEO technique
Cell voltage (V)	20-80	120-300
Current density (A/dm^2)	< 10	< 30
Substate pretreatment	Critical	Less critical
Common electrolytes	Sulfuric, chromic, or phosphoric	Neutral/alkaline (pH=7-12)
Coating thickness (μm)	< 10	< 200
Coating hardness	Moderate	Relatively high
Adhesion to substrate	Moderate	Very high
Temperature control	critical	Not so important

Table 1. General comparison between conventional direct current anodizing and plasma electrolytic oxidation coating technologies [12]

1.2. Process of PEO treatment

A typical equipment used for PEO treatment is shown in Figure 1. An enclosure (1) is mounted close to a high voltage AC power supply (2). The metal substrate to be PEO coated (3) is immersed in the electrolyte in a water cooled, insulated electrolyte tank (4) made of stainless steel and also serves as the counter electrode. The tank is insulated from ground and mounted in the safety interlocked enclosure (1), the latter being equipped with fume extraction facilities (5) and a window (6) to allow the PEO process to be viewed. The electrolyte is typically mixed (7) and recycled via a flow circuit containing a heat exchanger/chiller and a 50 to 100 mm filter (8) [12].

Before the PEO treatment, the samples should be ground and polished with abrasive paper, degreased ultrasonically in acetone and cleaned with distilled water. During the treatment, the samples are used as anode plates and immersed in the electrolyte which is cooled by a water cooling system and mechanically stirred by a mixer. After the treatment, the samples should be rinsed with distilled water and air dried [12].

1.3. Mechanism of PEO technique

When the samples of valve metals [13] or their alloys are placed in the electrolyte, the metal surface immediately generates a layer of insulating oxide film after the energization. The weak parts of the oxide film were broke down after the applied voltage exceeds a critical value,

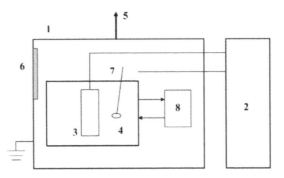

1. Insulated enclosure; 2. High voltage AC power supply; 3. Workpiece (light metal substrate); 4. Electrolyte holding tank and counter electrode (stainless steel); 5. Fume extraction vent; 6. Viewing window; 7. Electrolyte mixer; 8. Flow circulation via chiller and filter

Figure 1. Equipment used for PEO treatment [12].

resulting in the occurrence of micro-arc discharge phenomenon. The process of electrical breakdown of PEO involves many physical (such as crystallizing, melting, phase change at high temperature and electrophoresis, etc.), chemical (such as high-temperature chemistry, plasma chemistry) and electrochemical processes. A variety of models and hypotheses on the electrical breakdown of PEO process were created by many researchers to explain its causes. Therefore, the theory of electric breakdown also experienced different stages of development, such as thermal mechanism, mechanical effects and the mechanism of electron avalanche. The mechanism is so complex that, to date, no theory can give a complete and accurate explain of the whole PEO process. [14-15].

Recently, some studies have showed some basic processes of PEO treatment which involves: the formation of space charges in the oxide matrix; gas discharge generated in the pores of oxide; localized melting of the layer material; thermal diffusion; deposition of the colloidal particles; migration of negatively charged colloidal particles into the discharge channels; plasma chemical and thermochemical reactions, etc. [14-15].

1.4. Structures and compositions of PEO coatings

The PEO coatings generally consist of a porous top layer, compact intermediate layer and thin inner layer. The intermediate layer and inner layer are dense and adhered well to the substrate [12, 16]. The surface morphologies characterized with many micropores, microcracks and dimples [17]. Previous studies showed that the micropores were formed by molten oxide and gas bubbles thrown out of micro-arc discharge channels and the cracks resulted from the thermal stress due to the rapid solidification of the molten oxide in the relatively cold electrolyte.

The PEO ceramic coatings are composed of not only predominant substrate metal oxides (such as Al_2O_3, TiO_2 and MgO on Al, Ti, Mg and their alloys respectively)[12], but also more complex oxides and compounds which involve the components presented in the electrolytes (such as

Al_2O_3 on Ti6Al4V alloy in aluminate solution [18]; mullite on Al alloy in silicate solution [19]; MgF_2 on Mg alloy with KF in the electrolyte [20] and TiO_2 on Mg alloy in phosphate solution containing titania sol [21]).

1.5. Influence factors for PEO technique

It is considered that the PEO treatment is a multifactor-controlled process, which is influenced by many factors, intrinsic or extrinsic. The compositions of substrate materials and electrolyte are considered to be intrinsic factors which play crucial role for the structure and composition of PEO coatings, while the extrinsic factors generally consist of electrical parameters, processing temperature, oxidation time and additives [14]. Herein, these influence factors for PEO technique will be introduced and discussed briefly.

1.5.1. Influence of substrate materials

The difference of substrate materials plays a crucial role in the components and properties of PEO coatings. The predominant compositions of the PEO coatings depend on the substrate materials, for example, the main content of coatings deposited on Al, Mg, Ti and their alloys are Al_2O_3, MgO and TiO_2 respectively. Therefore, the PEO coatings deposited on different substrate materials generally possess different properties. According to recent studies, the available coating thicknesses are around 300 μm on Al alloy, 150 μm on Mg alloy and 200 μm on Ti alloy respectively. The hardness value ranges for PEO coatings formed on different substrate materials are generally from 300 HV to 2500 HV on Al alloy, from 200 HV to 1000 HV on Mg alloy and from 300 HV to 1100 HV on Ti alloy [14-15].

1.5.2. Influence of electrolytes

The compositions of electrolyte greatly affect the properties of PEO ceramic coatings. Different electrolytes result in different growth rates, structures, phase compositions and element distribution of the PEO coatings [22-25]. Generally, the electrolytes used for PEO treatment are composed of acidic electrolytes and alkaline electrolytes. The acidic electrolytes including concentrated sulfuric acid, phosphoric acid and other salt solutions etc. are seldom used at present due to their great environmental pollution. While the alkaline electrolytes mainly consist of four systems including sodium hydroxide based electrolytes, silicate based electrolytes, phosphate based electrolytes and aluminate based electrolytes[14].

1.5.3. Influence of electrical parameters

The whole PEO process and the properties of ceramic coatings are greatly affected by electrical parameters including current modes, current density, current frequency, anodic voltage, cathodic voltage, duty cycle etc. Recently, the effects of electrical parameters on PEO coatings were investigated by many researchers as follows.

The effect of electrode distance on anode current and the influence of anode current on PEO process were investigated by C.B. Wei et al [26]. The PEO processes were carried out on 2024 Al alloy in the electrolyte of sodium silicate with other additives, keeping the anode and cathode plates located face to face and separated by different distances. Results showed that the anode current was influenced by the distance between the electrodes and had a critical effect on the oxidation efficiency. The anode currents were found to decrease with larger distances. The current flowing through the front surface was higher than that through the back surface. The ball-on-disk tribological tests and corrosion tests revealed that the front surface has better tribological properties and higher corrosion resistance than the back surface.

R.O. Hussein et al. [27] studied the influence of different current modes on PEO plasma discharge behaviour and alumina coating microstructure. The PEO processes were carried out on pure 1100 aluminum using two different current modes of pulsed unipolar and bipolar in the electrolyte of Na_2SiO_3 (7 g/L) and KOH (1 g/L). It was found that the plasma temperatures vs. process time were different under different current modes. The plasma temperature spikes were believed to be caused by the strongest plasma discharges initiated at the interface between the oxide coating and substrate. Compared to the unipolar current process, the application of pulsed bipolar current resulted in reducing the high spikes on temperature profiles and the average plasma temperature. The aluminum oxide coating morphology and microstructure were also significantly different under different current modes. The bipolar current mode could improve the coating quality compared with the unipolar current mode, in terms of surface morphology and cross-sectional microstructure. A dense coating morphology could be achieved by adjusting positive to negative current ratio and their timing to eliminate or reduce the strongest plasma discharges and the high temperature spikes, thus resulting in the improvement of coating qualities.

Yue Yang et al. [28] investigated the effects of current frequency on microstructure and wear resistance of ceramic coatings embedded with SiC nanoparticles on Mg alloy produced by PEO. The PEO treatments were carried out on AZ91D Mg alloy in the electrolyte containing $NaAlO_2$ (20 g/L), NaOH (3 g/L) and SiC nanoparticles (2 g/L), with the current frequency fixed at 500 Hz, 700 Hz and 900 Hz respectively. Results revealed that with the increasing of current frequency, more SiC nanoparticles randomly dispersed on the surface. The thickness and growth rate of the coatings increased with the increasing of applied current frequency. Furthermore, the ceramic coating embedded with SiC nanoparticles formed at a current frequency of 900 Hz showed the finer microstructure, lower surface roughness and best wear resistance.

R.H.U. Khan et al. [29] studied the effects of current density on surface characterization of PEO treated Al alloy. The samples were fabricated by PEO process at different current density of 5, 10, 15 and 20 A/dm². It was found that the coating thickness increased with the increased current density. The largest coating thickness (40 μm) was obtained at 20 A/dm² current density, whereas a thinnest coating (3 μm) was formed at 5 A/dm². Furthermore, the relative content of α-Al_2O_3 tended to increase with increasing current density. Residual stresses in alumina coatings tended to decrease with the increased current density due to increased plasma microdischarge events which promoted stress relaxation through formation of microcrack network and thermal annealing in the coatings.

Ping Huang et al. [30] investigated the effects of different voltages on mechanical properties of titania prepared by PEO. The PEO treatment was carried out on pure titanium, in an aqueous solution containing calcium salt and phosphate salt, using different voltages from 240 V to 450 V. Results showed that the composition of the PEO coatings was generally anatase and rutile, while at higher voltage of 400-450 V, a new $CaTiO_3$ phase appeared. The pore size of PEO coatings increased with the increase of applied voltage. The samples prepared at 240-350 V had much stronger bonding strength compared to that prepared at higher voltage. The elastic modulus and residual stress both increased with the increasing of applied voltage.

The effects of cathode voltage on structure and properties of PEO ceramic coatings formed on NiTi alloy were investigated by F. Liu et al. [31]. The PEO processes were carried out on nearly equiatomic NiTi alloy in a solution of sodium aluminate and sodium hypophosphite, using a constant voltage mode, with anodic voltage kept constant at 400 V and cathodic voltage controlled at 0, 10, 20, and 30 V respectively. It was found that, the thickness and surface roughness of PEO coatings increased with the increasing of cathodic voltage, the friction coefficient of PEO coatings against GCr15 steel ball also increased, while the bonding strength of the coatings to the substrate and the corrosion resistance of PEO coatings both decreased. The PEO ceramic coatings formed at various cathodic voltages on NiTi alloy were composed of γ-Al_2O_3 at the only crystalline phase. The crystallinity could be enhanced through increasing the cathodic voltages. As a whole, the cathodic voltage applied for PEO showed a negative correlation with the biocompatibility of the ceramic coatings.

Yuming Tang et al. [32] studied the influences of duty cycle on the bonding strength of Mg alloy by PEO treatment. The duty cycle varied in the range of 10-40% with positive and negative cycle remained equal. It was revealed that the higher duty cycle increased the coating porosity and slightly decreased the thickness of the oxide coating. The PEO coatings mainly consisted of MgO and $MgSiO_3$. And the relative content of MgO in the coatings increased while the content of $MgSiO_3$ slightly decreased with the increase of duty cycle. Furthermore, as the duty cycle increased, the lap-shear strength of the bonding joints increased. The highest lap-shear strength (24.50 MPa) was obtained under the duty cycle of 40%. The reason was attributed to the larger porosity and enhanced mechanical interlocking effect.

1.5.4. Influence of processing temperature

The electrolyte temperature can greatly affect the PEO process. If the temperature is too low, the oxidation process becomes weak, resulting in less thickness and lower hardness of the PEO coatings. If the temperature is too high, the dissolution of oxide film will be enhanced, and thus cause the coating thickness and hardness to decrease significantly. Therefore, the processing temperature should also be studied and generally controlled in the range of 20-40°C.

H. Habazaki et al. [33] investigated the effects of different electrolyte temperatures on formation and characterization of wear-resistance of PEO coatings on Ti alloy. The PEO processes were carried out on Ti-15-3 alloy in the electrolyte of $K_2Al_2O_4$ (0.15 mol/L), Na_3PO_4 (0.02 mol/L) and NaOH (0.015 mol/L), at different electrolyte temperatures between 278 K and 313 K. Results showed that at the lowest temperature of 278 K, the yielded PEO coating contained higher concentration of α-Al_2O_3 phase in addition to the Al_2TiO_5 major phase, exhibited lower porosity, uniformity and density, and thus showed more improved wear resistance, compared to that formed at higher temperatures.

1.5.5. Influence of oxidation time

With the increasing of oxidation time, the coating thickness increases, while the growth rate decreases. Different oxidation time can result in different coating qualities, such as thickness, roughness, adhesion, hardness, wear resistance and corrosion resistance etc.. Therefore, the oxidation time for PEO treatment should be investigated and optimized.

Yanhong Gu et al. [34] studied the effect of oxidation time on corrosion behavior of PEO coatings on Mg alloy in simulated body fluid. The samples were fabricated on AZ31 Mg alloy in aqueous solution of sodium phosphate (30 g/L), using applied DC voltage of 325 V, current density of 150 mA/cm^2 and pulsed frequency of 3000 Hz, with 1, 3, 5 and 8 min different oxidation time. Results showed that the coatings mainly consisted of Mg, MgO, MgAl$_2$O$_4$ and Mg$_3$(PO$_4$)$_2$, and the oxidation time had very little influence on the phase compositions. The diameter of the micropores in the PEO coating surface increased with increasing oxidation time. The coating thickness increased with increasing oxidation time until 5 min (20 μm). The sample coated at 5 min showed the thickest layer with a relatively smooth and uniform microstructure with fewer micropores, compared to the other PEO oxidation times. When the oxidation time, however, was increased to 8 min, the coating thickness decreased and the coating surface became rough. The porosity decreased with increasing oxidation time until 5 min (4.40%), and then increased to 6.28% for an oxidation time of 8 min. As a whole, the PEO coating produced at 5 min had the smallest corrosion current density and the largest electro-chemical impedance, resulting in the highest corrosion resistance, due to the compact, smooth and uniform morphology of coating surface with lower porosity.

1.5.6. Influence of additives

Employing different additives in the electrolyte can greatly affect the PEO process, and thus resulting in different properties of the coatings. For example, Jun Liang et al. [20] studied the effect of KF in Na$_2$SiO$_3$-KOH electrolyte on the structure and properties of PEO coatings formed on Mg alloy. It was found that the addition of KF contributed to increase the electrolyte conductivity, decrease the work voltage and final voltage in the PEO process and change the spark discharge characteristics. Furthermore, the addition of KF resulted in a decrease of pore diameter and surface roughness, an increase of the coating compactness and the changes in the phase compositions as well. The hardness and wear-resistance of the coating also enhanced due to the addition of KF.

2. Tribological properties of PEO coatings

Employing PEO technique to form ceramic oxide coatings on Ti, Mg, Al and their alloys can significantly enhance the mechanical and tribological properties, such as high hardness, superior wear resistance and good adhesion to the substrate. In recent years, investigations on the phase composition, mechanical and tribological properties of PEO coatings on Ti, Mg, Al and their alloys were done by many researchers. However, the tribological performances of PEO coatings are not only affected by the intrinsic properties of PEO coatings, but also affected by many extrinsic factors, such as sliding loads, sliding speed, counterpart materials, lubricated conditions, temperature and humidity etc. Herein, the sliding loads are emphasized and classified into three levels: low loads (0-5 N), medium loads (5-50 N) and heavy loads (above 50 N). And then, the friction and wear behaviors of different PEO coatings in different conditions will be introduced and discussed under different load levels.

2.1. Friction and wear behavior of PEO coatings under low loads

The microstructure, mechanical and tribological properties of PEO coatings formed on Ti6Al4V alloy were studied by Y.M. Wang et al [35]. A nanoindentation test showed that the hardness and elastic modulus were about 8.5 GPa and 87.4 GPa for the compact region of the PEO coating, and about 4 GPa and 150 GPa for the Ti6Al4V substrate. The hardness and elastic modulus were mainly constant in the compact region within 33 μm, and decreased remarkably beyond 33 μm to the outer surface. A sheer test showed that the adhesion strength between coating and substrate was about 70 MPa. The tribological behaviors of untreated and PEO coated samples were evaluated by a pin-on-disk tribometer under the normal loads of 0.3, 0.5, and 1 N, with a sliding speed of 0.05 m/s or 0.15 m/s, using SAE52100 steel ball as counterpart material. Results of friction and wear tests showed that the friction coefficient of PEO coating against steel was as low as 0.2-0.3 at loads not more than 1 N and sliding cycles within 2500 times, and gradually increased at the later stage of wear test due to the oxidation and materials transfer wear mechanism.

The investigations of structure, composition, mechanical and tribological properties of PEO coatings formed on AM60B Mg alloy in silicate and phosphate electrolyte have been done by Jun Liang et al. [36] The samples were fabricated in the electrolyte containing Na_2SiO_3 (10 g/L), KOH (1 g/L) or Na_3PO_4 (10 g/L), KOH (1 g/L). The coating formed in silicate electrolyte is composed of periclase MgO and forsterite Mg_2SiO_4 phases while MgO and a little of spinel $MgAl_2O_4$ are the main phases of the coating formed in phosphate electrolyte. Generally, the forsterite Mg_2SiO_4 has a greater hardness than that of the MgO. Therefore, the coating formed in silicate electrolyte exhibits a higher microhardness than that formed in phosphate electrolyte. The friction and wear properties of the PEO coatings were evaluated on a reciprocal-sliding UMT-2MT tribometer in dry sliding conditions under a load of 2 N, using Si_3N_4 ball as counterpart material, with a siding speed of 0.1 m/s and sliding amplitude of 5 mm. The wear life of PEO coatings formed in two different electrolytes was compared with the thin coatings and results showed that the wear life of coating formed in silicate electrolyte is about four times as long as that of coating formed in phosphate electrolyte. The uncoated Mg alloy has a friction coefficient of about 0.3 and exhibits a high wear rate of 3.81×10^{-4} mm^3/Nm. While for both the oxide coatings, the friction coefficients are in the range of 0.6-0.8 and the wear rates are only in the range of 3.55-8.65×10^{-5} mm^3/Nm. These evidences demonstrate that the PEO coatings formed on Mg alloy in both electrolytes have greatly enhanced the wear resistance but exhibit higher friction coefficients compared with the uncoated Mg alloy. Furthermore, the oxide coating formed in silicate electrolyte has a higher friction coefficient but exhibit a better wear resistance than that formed in phosphate electrolyte. It also suggests that the structure and phase composition of coatings are indeed the dominant factors which influence the mechanical property and friction and wear behaviors of PEO coatings.

2.2. Friction and wear behavior of PEO coatings under medium loads

P. Bala Srinivasan et al. [37] studied the dry sliding wear behaviour of PEO coatings with different thickness of 10 μm and 20 μm on cast AZ91 magnesium alloy. The samples were fabricated by PEO treatment in silicate based electrolyte containing Na_2SiO_3 (10 g/L) and KOH (10 g/L). The dry sliding wear behaviour of the untreated Mg alloy, PEO coated specimen A and B was assessed on a ball-on-disc oscillating tribometer, under three different loads of 2N, 5N and 10N, with an oscillating amplitude of 10 mm and at a sliding velocity of 5 mm/s for a sliding distance of 12 m, using an AISI 52100 steel ball of 6 mm diameter as static friction partner. For the uncoated Mg alloy, the friction coefficients were fluctuating in the range of 0.24-0.40 under different loads. For the 10 μm PEO coating, the friction coefficient reached to a steady value of about 0.78 under 2 N load, dropped to around 0.35 after a sliding distance of about 4 m under 5 N load, while dropped in a very short time under 10 N load. For the 20 μm PEO coating, the friction coefficients did not drop at all loads and remains steady. Moreover, the friction coefficient showed lower value with an increase in load (0.8 at 2 N, 0.68 at 5 N and 0.62 at 10 N). The uncoated Mg alloy under all loads and the 10 μm PEO coating under 5 N and 10 N loads all showed high wear rates. While the 20 μm PEO coating under all loads and the 10 μm PEO coating under 2 N load all showed much lower wear rates. The results indicated that the thickness of coatings played a crucial role in enhancing the wear resistance. At higher initial stress levels, the deformation of the substrate causes the cracking and flaking-off of the coating, especially when it is thin. Under such circumstances the increased thickness of PEO coating provided a better load bearing capacity, thus resulting in a superior wear resistance.

M. Treviño et al. [38] investigated the wear of coatings on Al 6061 alloy fabricated by PEO treatment in Na_2SiO_3-KOH electrolyte. The coatings with different thickness of 100, 125 and 150 μm were fabricated and characterized. Composition analysis showed that the coatings consisted of a combination of oxide phases such as mullite, α-Al_2O_3, γ-Al_2O_3 and amorphous alumina. It was suggested that the presence of α-Al_2O_3 phase presented the greatest wear resistance compared with other phases such as mullite, γ-Al_2O_3 and amorphous alumina which were highly vulnerable for the conditions studied. No difference was detected for the different coatings in hardness values which were of 1556±11 HV_{50} compared with that of 109±3 HV_{50} for the substrate. The tribological properties were evaluated by a pin on disc test machine, with a sliding distance of 1 km and a constant linear speed of 13.76 m/min, using 10, 20, 30 and 40 N different normal loads for each coating thickness. Friction and wear tests showed that the friction coefficient changed along the tests, and the weight loss depended on both the thickness of the coatings and the loads applied during the test. The wear mechanisms were suggested to be adhesion and abrasion by hard particles. The thinnest coating of 100 μm exhibited better resistance to wear and showed the friction coefficients which exhibited a continuous increase independently of the applied loads. The friction coefficients for the coating of 125 μm remained constant when loads of 10 and 20 N were used, and reduced their values once a certain distance was achieved when tested with 30 and 40 N. The friction coefficients for the coating of 150 μm were found to increase under a load of 10 N, to remain fairly constant with loads of 20 and 30 N and to reduce their value once a distance of around 500 m was achieved with a load of 40 N. The reduction of friction coefficients for the coatings of 125 μm and 150 μm suggested that the coatings were completely removed under the loads of 30 N and 40 N resulting in contact with the alloy substrate which was probably lubricated by wear debris generated.

2.3. Friction and wear behavior of PEO coatings under heavy loads

Chen Fei et al. [39] studied the tribological performance of PEO ceramic coatings fabricated on Ti6Al4V alloy in the electrolyte containing Na_2SiO_3 (10g/L), Na_2CO_3 (4g/L) and EDTA-2Na (5g/L). Coatings with a thickness of 10 μm were formed and polished to remove the prominent ceramic particles of the outer surface in order to reduce the effect of roughness on tribological behavior. The tribological behaviors of unpolished coating, polished coating and untreated Ti6Al4V alloy were evaluated on a ball-on-disk tribometer under the dry sliding conditions, using balls of SAE52100 steel as counterpart materials, with normal load of 100 N, rotation speed of 1000 rpm, sliding speed of 0.42 m/s and sliding time of 10 min. For the untreated Ti6Al4V alloy, the long-term friction coefficient is about 0.4, and the worn surface that sliding against steel revealed that the dominant wear mechanism is extensive abrasive and adhesive wear. For the unpolished PEO coating, the friction coefficient exhibited a high value of about 0.5, and the wear track showed severe abrasive wear, also accompanied by severe adhesive wear from the steel counter surface leading to material transfer on the coated surface. The porous surface of the unpolished PEO coatings is very rough due to the scraggy ceramic products. Unlike sliding that usually leads to plastic shearing in materials, the impact caused by the ceramic asperities on the surface results in catastrophic failure, such as cracking and crushing of the contact regions, which leads to faster material removal and the production of the sharp ceramic debris fragments. In contrast, for the polished coating, the friction coefficient exhibited a relatively low and stable value, almost remaining constant at 0.2. As the outer surface was polished to remove the prominent ceramic particles, the initial contact conditions were changed from a rough ceramic/steel to a smooth ceramic/steel mating surface. Therefore, the cracking and crushing of prominent ceramic regions due to great vibrations were eliminated. Results showed that the worn surface was relatively smooth, accompanied with fine debris embedded in the edges of contact regions. The good antifriction properties are attributed to the microstructure of the coatings which are mainly composed of rutile and anatase TiO_2. TiO_2 especially the rutile-type, is known as a potentially low friction and wear reducing material.

Jun Tian et al. [40] investigated the structure and antiwear behavior of PEO coatings on 2A12 Al alloy. The samples were fabricated by PEO treatment in the electrolyte composed of Na_2SiO_3 (30g/L), NaOH (5g/L), with current density controlled to below 10^3 A/m^2. The as-deposited coatings were polished with SiC paper to remove 20%, 30%, 40% and 50% of the whole thickness of the coatings as polished coating samples. The results of structural and phase composition analysis showed that the PEO coatings on Al alloys showed two distinct layers, i.e. a porous outer layer consisting predominantly of γ-Al_2O_3 and a dense inner layer consisting predominantly of α-Al_2O_3. The inner layer α-Al_2O_3 has better antiwear ability compared with the outer layer γ-Al_2O_3. Therefore, with the increasing of the coating thickness, the antiwear life of the outer layer becomes smaller than that of the inner layer. The results of friction and wear tests showed that the polished coating mainly composed of α-Al_2O_3 registered a lower wear rate of 3.00-5.00×10^{-6} mm^3/Nm in reciprocating sliding against ceramic counterpart at a speed of 0.33 m/s and a contact pressure of 2 MPa. The antiwear life of the polished coating reached 2500 m at a speed of 1.25 m/s and a load of 300 N, and the friction coefficient was more than 0.45 against the steel ring in a Timken tester which was a little lower than that of the out layer registering more than 0.47.

The aforementioned studies revealed that the PEO ceramic coatings can sharply increase the wear resistance and decrease the wear rate, compared to the uncoated substrates. However, the PEO coatings normally exhibit higher friction coefficients which can cause not only the wear of sliders, but also the wear damage of counterpart materials in many tribological applications. Thus, it is necessary to fabricate the PEO coatings with both good wear resistance and low friction coefficient.

3. Improvements of tribological behavior of PEO coatings

In order to further improve the tribological properties of the PEO-treated lightweight metals, many attempts to reduce the friction coefficient of the PEO coatings have been made. Herein, three main developments in improvement of tribological properties of PEO coatings are reviewed, which can be categorized as (1) liquid lubrication, (2) duplex coatings and (3) composite coatings.

3.1. Liquid lubrication for improving the tribological behavior of PEO coatings

As there are many micropores, microcracks and dimples on the surface of the PEO coatings [17], these pores, cracks and dimples can act as reservoirs for oil lubricants, which may result in a positive effect to the tribological performance of PEO coatings under boundary-lubricated conditions.

Studies on the wear resistance of PEO coatings on 2024 Al alloy under oil-lubricated condition were done by Tongbo Wei et al. [41]. The friction and wear tests were carried on an MRH-3 ring-on-block tester, at a ring linear speed of 2.60 m/s and normal loads from 300 N up to 1400 N, using AISI-C-52100 steel rings and aluminum rings covered with polished PEO coatings as counterpart. Commercial 4838 lubricating oil was used as the lubricating medium. Friction and wear test showed that the friction coefficient of polished coatings was within 0.020-0.060 under oil-lubricated condition which was reduced to about 1/10 compared with that under dry sliding condition registering within 0.20-0.35, and the wear rate of polished coating was within $1.00\text{-}8.50\times10^{-9}$ mm^3/Nm which was reduced to be about 1/1000 compared with that under dry sliding condition registering within $1.00\text{-}2.00\times10^{-6}$ mm^3/Nm. The polished coatings showed excellent wear-resistance in oil-lubricated sliding against steel and Al$_2$O$_3$ ceramic ring and can endure a sliding distance as large as 18.7 km at loads as high as 1400 N.

Fei Zhou et al. [42] investigated the friction characteristic of PEO coating on 2024 Al alloy, sliding against Si$_3$N$_4$ balls, in water and oil environments, at different normal loads and sliding speeds. Results showed that, with the increasing of normal load and sliding speed, the friction coefficient of the PEO/Si$_3$N$_4$ tribopair in water and oil decreased from 0.72 to 0.57 and 0.24 to

0.11 respectively. The wear mechanism of the PEO coatings changed from abrasive wear in air to mix wear in water, and finally became microploughing wear in oil.

M.H. Zhu et al. [43] investigated the fretting wear behaviors of PEO coating on LD11 Al alloy sealed by grease. It was found that the friction coefficient of the sealed PEO coating under all test parameters were greatly lower than that of the PEO coating. At the same time, there was a longer stage with low friction coefficient that can be observed in the friction coefficient curves for all test conditions of the sealed PEO coating. It was clear that the sealed PEO coating presented an obvious lubricating action during the fretting wear processes. In partial slip regime, the damage of the two coatings was very slight, and the porous structure was still intact even after 104 cycles. The fretting wear mechanisms of the two coatings in slip regime were main abrasive wear and delamination, but higher proportion of the traces of relative sliding was presented on the scars of the sealed PEO coating. As a conclusion, the sealed PEO coating exhibited a better resistance for alleviating fretting wear and lengthening service life than that of the PEO coating.

The fretting wear behaviour of PEO coatings formed on Ti6Al4V alloy under oil lubricated conditions was studied by Yaming Wang et al. [44]. The fretting wear tests of PEO coatings were conducted on a PLINT fretting fatigue machine under unlubricated and oil lubricated conditions (smear and oil bath lubrication), using 52100 steel ball as vibrated counterpart material and with a small reciprocating amplitude of 60 μm. The results showed that in unlubricated condition, the friction coefficient rapidly increased up to 0.8-0.9 and maintained relatively stable. In smear oil lubricated condition, the friction coefficient showed an obvious higher value within 0.18-0.43 in the range of about 2500-6700 cycles. While in oil bath lubricated condition, the friction coefficient reduced significantly to a low and stable value of 0.15 in the long-term fretting test. This indicated that the coatings with oil lubrication lowered the shear and adhesive stresses between contact surfaces, and consequently alleviated the possibility of initiation and propagation of cracks in the inner layer of the coating or titanium alloy substrate.

3.2. Duplex coatings for improving the tribological behavior of PEO coatings

Employing liquid lubricants may improve the tribological properties of the PEO coatings. While in rigid and severe working conditions, such as high vacuum, high temperature, chemical and radioactive environments, liquid lubricants often do not function [45]. Furthermore, liquid lubricants may contaminate the workpieces. Therefore, some multi-step preparation methods combined with the PEO process are employed to fabricate PEO-based duplex coatings on the metallic substrates. The duplex coatings are formed by one of the post treatments (mainly including impregnation, spraying and chemical/physical vapour deposition) on the yielded PEO ceramic coatings. These duplex coatings can sharply decrease the friction coefficient and improve the wear resistance. Herein, some successful applications for the duplex coatings were introduced briefly.

3.2.1. Impregnation

Because the PEO ceramic coating is formed on the metal surface via a series of localized electrical discharge events, there are many micropores left in the coating [17]. This provides the probability to deposit small sized solid lubricant particles into these micropores to form a binary coating [46]. Herein, a simple and effective method of vacuum impregnation was introduced. In this method, the PEO coating samples are immersed into water-based solid lubricant suspension, and put into a vacuum oven. When deposited for a set time, the samples are heated for a period of time in high temperature for solidification. The solid lubricant particles can impregnate into the micropores of the PEO ceramic coating under the vacuum. With the increasing of deposition time and heat treatment, a compact top film covering the PEO ceramic coating is formed for anti-friction purpose.

Zhijiang Wang et al. [47] investigated the properties of a self-lubricating Al_2O_3/PTFE duplex coating formed on LY12 Al alloy by PEO treatment combined with vacuum impregnation of PTFE. In their work, the PEO coating samples were immersed into water-based PTFE suspension, and putted into a vacuum oven (less than 5×10^{-3} torr). When deposited for a set time, the samples were heated for 24 h at 200℃. Results showed that the PTFE powder particles with the size in the range of 100-170 nm could deposit into the PEO ceramic coating and covered the rough and porous surface. Tribological tests showed that the friction coefficient and wear mass loss of the Al_2O_3/PTFE duplex coating decreased sharply. As the cracks and micropores of the PEO coating were filled by solid lubricant, PTFE could form a lubricating film on the frication surface of the steel ball when the steel ball which worked as counter material slided against the coating. With increased sliding distances, the solid lubricant provided continuous supply due to the abundance of PTFE lying in the micropores of the PEO coating. At the same time, the PEO coating could play the role as the wear-resistant substrate to support soft PTFE polymer. As a result, the friction coefficient of the self-lubricating coating could remain at a constant with minimal weight loss during the long-term sliding.

3.2.2. Spraying top coatings

The porous feature of the PEO coating opens a good way to introduce solid lubricant into micropores or depositing on the surface of coating. And the presence of pores affords an effective mechanical keying between solid lubricant topcoat and the PEO layer. Spraying is a simple, effective and low cost method as a post treatment to apply solid lubricant on PEO coating to form a self-lubricating duplex coating. The PEO ceramic coating serves as underlying loading layer and solid lubricant top layer plays the roles as friction reducing agent.

Y.M. Wang et al. [48-49] have successfully prepared a self-lubricating duplex coating on Ti6Al4V alloy by PEO treatment combined with spraying graphite process. The spraying graphite process forced on the surface of the PEO coating was carried out using a self-made spraying gun with 4 atmosphere pressure, followed by solidification at 180℃ for 15 min. Results showed that the surface of PEO coating was characterized by micropores of different size and shape and covered by graphite lubricant exhibiting a special shape of plate. The duplex coating exhibited good antifriction property, registering friction coefficient of about 0.12, which is 5 times lower than that of the PEO coating sliding in the similar condition.

C. Martini et al. [50] have successfully fabricated a self-lubricating duplex coating on Ti6Al4V alloy by PEO treatment combined with spraying PTFE process. The PTFE topcoat deposited by spraying a solvent-based aerosol suspension proves to be beneficial in terms of both friction and wear resistance, particularly in an intermediate (30-50N) load range. The friction coefficient of duplex coating reduced from 0.8-1 to 0.2-0.3, which is attributed to the anti-friction properties of PTFE.

3.2.3. CVD/PVD to form top films

Chemical vapour deposition (CVD) and physical vapour deposition (PVD) techniques are well known to deposit hard coatings such as TiN, CrN and DLC etc., for providing surfaces with enhanced tribological properties in terms of low friction coefficient and high wear resistance. In recent years, attempts have been made to introduce CVD or PVD coatings on components of machines and engines. However, in practice TiN-, CrN- or DLC-coatings on alloys of light metals formed by various CVD/PVD methods often exhibit limited tribological performance due to the elastic and plastic deformation of the substrates under mechanical loadings, which can result in eventual coating failure, since the coatings are usually too thin to support the heavy loads and protect the substrates in the contact conditions [51]. Deposition of thick (e.g. >10 μm) CVD/PVD coatings usually results in high compressive stresses, and thus low adhesion [52].

Employing PEO technique can deposit thick ceramic coatings which exhibit high hardness, superior wear resistance and excellent load-bearing capacity. However, these PEO coatings generally exhibit high friction coefficients which can limit the wear resistance and cause the wear damage of counterpart materials. Therefore, a multi-step preparation of duplex coatings by PEO and CVD/PVD can integrate the advantages of excellent load-bearing capacity of PEO coatings and low friction coefficient of CVD/PVD coatings. In recent years, successful applications of these methods were done by some researchers as follows.

Samir H. Awad et al. [53] studied the tribological properties of duplex Al_2O_3/TiN coatings on 2A12 Al alloys deposited by a combined plasma electrolytic oxidation (PEO) and arc ion plating (AIP) technique. The thickness of the Al_2O_3 coatings and TiN coatings were 30-40 μm and 3-5 μm, respectively. The tribological properties were evaluated by ring-on-ring tests at speeds of 0.75 and 1.25 m/s, under loads of 98, 300, 500, and 800 N, using a GCr15 bearing steel ring as rotated counterpart material. Results showed that the duplex coatings possessed very high hardness and wear resistance, and their mechanical and tribological properties were better than those of single TiN coatings, single PEO coatings and the uncoated Al alloy substrate. The Al2O3 intermediate layer played a crucial role in providing the load support essential to withstanding sliding wear at high contact loads.

X. Nie et al. [54] investigated the tribological performances of duplex Al_2O_3/DLC coatings on Al alloy fabricated by a combined PEO and PI^3 (plasma immersion ion implantation) technique. The alumina ceramic coatings with a thickness of 50-60 μm were formed on BS Al-6082 aluminum alloy by PEO treatment and DLC coatings with 2-5 μm thickness were deposited on top of the PEO coatings. All the duplex alumina/DLC coatings exhibited a hardness of over 2000 HK_{10g}. The tribological properties of Al alloy, alumina coating on Al alloy, DLC coating on Al alloy and duplex alumina/DLC coatings on Al alloy were evaluated by pin-on-disc tribological tests, under a 10 N normal load, 0.1 m/s sliding speed and 50% RH, using SAE 52100 bearing steel (BS) or WC-Co (WC) balls as counterpart materials. Both the untreated Al alloy substrate and the PEO alumina coating gave a high friction coefficient of above 0.7 against both counterpart materials. Both the alumina and alumina/DLC duplex coatings exhibited excellent wear resistance, registering the wear rates in a low range of $1.4\text{-}1.9\times10^{-6}$ mm^3/Nm siding against WC-Co. Only the alumina/DLC duplex coatings provided a low and stable friction coefficient in a low range of 0.1-0.22. Whereas the single DLC coating on Al alloy failed quickly only at 25 m sliding distance due to its low load bearing capacity.

The tribological properties of duplex PEO/DLC coatings on Mg alloy formed by a combined plasma electrolytic oxidation and filtered cathode arc deposition technique were investigated by Jun Liang et al. [55]. The DLC film deposited on PEO coating was not uniform due to surface roughness of interface. The friction and wear properties of the uncoated Mg alloy, the polished PEO coating, single DLC film on Mg alloy and the duplex PEO/DLC coating were evaluated on a reciprocating ball-on-disk UMT-2MT tribometer, under a load of 2 N, with a sliding speed of 0.1 m/s and sliding amplitude of 5 mm, using Si$_3$N$_4$ ball as counterpart material. For the uncoated Mg alloy substrate, the friction coefficient varied in the range of 0.2-0.4 accompanied by severe oscillation. Severe wear and seizing were observed. Even for the DLC deposited Mg alloy substrate, the tribological behavior did not improve significantly. The DLC film failed quickly at around 350 s after starting the sliding test, due to the low load-bearing capacity of soft substrates. The polished PEO coatings showed a very high friction coefficient of around 0.7-0.8. But the deposition of DLC film on PEO coatings improved the dry friction behaviors significantly. The friction coefficient of the duplex PEO/DLC coating remained steady and less than 0.2, independently of the nonuniformity of coating surface.

E. Arslan et al. [56] studied the tribological performances of duplex titania/DLC coatings deposited on Ti6Al4V alloy using combined PEO and CFUBMS (closed field unbalanced magnetron sputtering). The thickness of PEO coating and DLC coating were about 10 μm and 6 μm respectively. The tribological properties were evaluated by a pin-on-disk tribometer (Teer-POD2), under a load of 2 N, at a speed of 100 rpm, a relative humidity of 45% and room temperature, using Al$_2$O$_3$ balls as counterpart materials. For the PEO coatings, the friction coefficient was considerably high, approximately above 0.45, and fluctuated, and the wear tracks were quite broad and rough with debris. For the single DLC coatings, the friction coefficient increased abruptly after 420 seconds due to severe failure caused by their low load bearing capacity. While for the duplex PEO/DLC coatings, the friction coefficients were much more stable, and approximately at 0.1. These results indicated that the duplex PEO/DLC coatings exhibited better tribological behaviors than the PEO and DLC coatings deposited on lightweight metals substrate.

3.3. Composite coatings for improving the tribological behavior of PEO coatings

The duplex coatings fabricated by the aforementioned methods can sharply decrease the friction coefficient and wear rate. However, these coatings are generally comprised of two layers: an inner PEO ceramic layer and an outside solid lubricant layer. When the outside layer is worn through, the tribological properties will be back to its original level. M. Aliofkhazraei et al. have successfully fabricated TiO$_2$/Al$_2$O$_3$ [57] and TiO$_2$/Si$_3$N$_4$ [58-59] nanocomposite coatings by PEO treatment on pure titanium. The PEO processes were carried out in the electrolyte of sodium-silicate (15 g/L), sodium-phosphate (2 g/L) and potassium hydroxide, adding Al_2O_3 or Si_3N_4 fine nanopowder. It was found that due to the incorporation of Al_2O_3 or Si_3N_4 nanoparticles in the coatings, the wear resistance and hardness of TiO_2/Al_2O_3 and TiO_2/Si_3N_4 nanocomposite coatings both increased significantly. However, the friction coefficients were still high or even higher, which could easily cause the wear damage of counterpart materials.

Recently, an alternative approach to obtain PEO coatings with low friction property was to introduce low friction materials into the coating by modifying the electrolytes with the solid lubricants additives. In this approach, solid lubricant particles, such as graphite, PTFE, MoS_2, WS_2 etc. are added into the electrolyte and dispersed with mechanical stirring to form a suspension. During the PEO process, solid lubricant particles can move from the electrolyte to the surface of the specimen, and be adsorbed on the surface, then be embedded into the ceramic coating.

It is important for this approach that solid lubricant particles should be sufficiently and uniformly dispersed in the electrolyte. So sufficient and constant mechanical stirring is inevitable. What's more, if necessary, a dispersant (such as acetone, ethanol etc.) is used to wet and disperse the solid lubricant particles. A kind of anionic surfactant (e.g. Sodium dodecyl sulfonate, etc.) is also used as additive to help the solid lubricant particles to be negatively charged and be suspended in the electrolyte. But the quantity of added dispersant and surfactant should be controlled and optimized. Too much additive can greatly affect the original properties of electrolyte and the whole coating process, resulting in low qualities of the coatings, such as nonuniformity, high roughness, poor adhesion to the substrate, less thickness, more inclination to breakdown and burn out, etc. However, lower concentration of additive can't wet and disperse the solid lubricant particles in the electrolyte sufficiently. So the specific and accurate quantity of additives should be decided by different coating processes.

It is generally considered that the embedding of solid lubricant particles into the ceramic coating matrix depends on concentration diffusion and electrophoretic deposition. The embedding of particles may be recognized by the adsorption of particles on the surface of the specimen, so higher concentration can help to enhance the adsorption rate, thus lead to more particles embedded into the ceramic coating. To be negatively charged are beneficial to the electrophoresis of particles in the electrolyte, thus resulting in more particles incorporated into the ceramic coatings. On the other side, the concentration of solid lubricant particles in the electrolyte and the quantity of solid lubricant particles incorporated into the coating should also be controlled and optimized. Too higher concentration of particles in the solution may greatly affect the original properties of electrolyte, causing poor qualities of the coatings. Too much solid lubricant particles incorporated into ceramic coatings may cause the destruction of the original coating structure and less ceramic component which plays the role as wear-resistant substrate, resulting in poorer qualities and lower wear resistance of the coatings.

It is also considered that the embedding of nanoparticles into the PEO coatings depends on current density, frequency, duty cycle and coating time. M. Aliofkhazraei et al. [58-59] investigated the effects of concentration, current density, frequency, duty cycle and coating time on the embedding of Si_3N_4 nanoparticles into TiO_2 coatings. Results showed that the wear mass loss rate decreased with the increasing of relative content of Si_3N_4 in the coatings. And the relative content of Si_3N_4 in the coatings increased by increasing of concentration, frequency and coating time, while it decreased by increasing of duty cycle and current density.

Up to now, some researchers have successfully incorporated solid lubricant particles (such as graphite, PTFE, MoS_2, etc.) into the PEO ceramic coatings formed on Al, Mg and Ti alloys. In the friction and wear process, the ceramic oxide coating plays the role as wear-resistant substrate while solid lubricant particles act as friction reducing agent during the sliding. Compared with the single PEO coatings, the yielded self-lubricating composite coatings can sharply decrease the friction coefficient and wear loss during the long-term sliding. Furthermore, the wear damage of counterpart materials can also be reduced greatly due to lower friction coefficient.

Xiaohong Wu et al. [60] have successfully incorporated graphite into Al_2O_3 ceramic coating fabricated on 2024Al alloy by PEO technique in a graphite-dispersed sodium aluminate electrolyte. The thickness of the composite coating produced was in the range 22±1 μm. Ball-on-disk tribological tests showed that the self-lubricating composite coating formed in the electrolyte containing 4g/L graphite exhibited a lowest friction coefficient of 0.09, under a normal load of 1 N, with a sliding time of 8 min and linear sliding speed of 0.08m/s, using a ball of Si_3N_4 as counterpart material.

Ming Mu et al. [61] have also successfully incorporated graphite into TiO_2 ceramic coating fabricated on Ti6Al4V alloy by PEO technique in a graphite-dispersed phosphate electrolyte. The tribological evaluation was carried out on a ball-on-flat UMT-2MT tribometer, under a constant normal load of 2N, with a frequency of 5 Hz, an oscillating amplitude of 5 mm and a sliding time of 30 min, using AISI52100 steel balls as counterpart materials. And the results of friction and wear tests showed that the friction coefficient of the PEO coating reduced from nearly 0.8 to about 0.15 and the wear resistance improved significantly under dry sliding conditions, due to the presence of the graphite particles in the coating. The specific wear rate of PEO/graphite composite coating decreased significantly, which registered to be around $8.6×10^{-6}$ mm³/Nm, compared with that of the uncoated alloy and pure PEO coating, which registered to be $5.2×10^{-5}$ mm³/Nm and $1.7×10^{-5}$ mm³/Nm respectively. The worn surface of uncoated alloy showed typical features of abrasive and adhesive wear, which resulted in high wear rate. For the pure PEO coating, the main forms of wear damage were suggested to be abrasive wear and detachment of the TiO_2 coating which made its friction coefficient and wear rate high. While for the PEO/graphite composite coating, the worn surface appeared quite smooth and showed no evidence of appreciable detachment of the coating. It was deduced that the graphite particles in the coating could be exposed to the wear track and then smeared on the contact surfaces which acted as solid lubricants in dry sliding wear condition, making the friction coefficient low throughout the sliding test. And consequently, the abrasive wear and detachment of the coating was effectively reduced.

Jie Guo et al. [62] tried to introduce PTFE nanoparticles suspension into the electrolyte to fabricated a PTFE-containing multifunctional PEO composite coating on AM60B Mg alloy. The samples were fabricated by PEO treatment in the electrolyte containing Na_3PO_4 (10.0 g/L),

KOH (1.0 g/L), with the addition of 3 vol.% PTFE nanoparticles suspension (10 wt%). In the PTFE-dispersed suspension, a nonionic surfactant (octylphenol polyoxyethylene ether, with the addition of 1-2 vol.%) and an anionic surfactant (sodium dodecyl sulfonate, with the addition of 2-4 vol.%) were used for PTFE nanoparticles dispersion and surface charge adjustment. Results showed that such PTFE-containing composite coating exhibited superior corrosion resistance, excellent self-lubricating property and better hydrophobic property when compared with pure PEO coatings. The PTFE-containing PEO coating exhibited a low and stable friction coefficient of less than 0.2 and low wear rate.

Recently, Ming Mu et al. [63] have once again successfully incorporated MoS_2 into TiO_2 ceramic coating fabricated on Ti6Al4V alloy by PEO technique in MoS_2-dispersed phosphate electrolyte. The electrolyte was prepared using Na_3PO_4 (20.0 g/L), KOH (2.0 g/L) in distilled water, with addition of MoS_2 particles (20.0 g/L), ethanol (100 ml/L) and an additive (0.5 g/L). Results showed that the TiO_2/MoS_2 composite coating exhibited improved tribological properties compared with the TiO_2 coating under dry sliding condition, which reduced the friction coefficient from 0.8 to about 0.12 and decreased the wear rate from 1.7×10^{-5} mm^3/Nm to 5.5×10^{-6} mm^3/Nm. It also should be noted that the TiO_2/MoS_2 composite coating showed better tribological property than the PEO/graphite composite coating under the same conditions.

From above studies, it is clear that the approach to prepare self-lubricating composite coating was much more effective than the duplex approaches in practice, for the PEO coatings contained low friction materials could be obtained by only one step. Besides, the coatings were expected to integrate the advantages of wear resistance of the PEO coating and low friction property of solid lubricants.

4. Concluding remarks

Plasma electrolytic oxidation (PEO) is a relatively novel technique that can be used to form metallurgically bonded ceramic coatings on some valve metals [13], such as Ti, Mg, Al, Nb, Zr and Ta etc. PEO process is now widely used to improve the surface performances of nonferrous metals by virtue of its high effectiveness, more convenience, economic efficiency and environmental performance. Moreover, the PEO ceramic oxide coatings deposited on light metals and their alloys generally exhibit superior wear resistance, large thickness, high microhardness and good adhesion to substrates. Therefore, in many tribological applications, employing PEO treatment to deposit ceramic coatings on surface of light metals can greatly improve the wear resistance, decrease the wear rate, enhance the wear life and reduce the wear damage of workpieces, thus resulting in good economic efficiency. However, it is necessary to realize the following conclusions about the tribological properties of PEO coatings.

Firstly, the thickness of the coatings plays a crucial role to possess a better wear resistance. For at high stress levels, the deformation of the substrate under loading/sliding can cause the cracking and flaking-off of the thin coatings, and thus result in severe failure. So higher thickness of PEO coating can provide a better load bearing capacity and thus possess superior wear resistance and enhanced wear life.

Secondly, the components of the PEO coating also greatly affect the friction and wear performances of the coating. It is found that the relative content of α-Al_2O_3 phase in PEO coatings on Al alloy, Mg_2SiO_4 phase in PEO coatings on Mg alloy and rutile phase in PEO coatings on Ti alloy plays a crucial role in presenting higher wear resistance of PEO coatings.

Thirdly, as the PEO ceramic coatings generally consist of a porous outer layer and a compact inner layer, in many tribological applications, the PEO coatings are polished with abrasive papers to remove the porous outer layer and get a higher hardness and a lower roughness. It is found that the polished PEO coatings generally exhibit improved friction and wear behaviors than the original PEO coatings and the untreated alloy substrates.

And last but not least, although the wear resistance has significantly enhanced, the PEO coatings deposited on the alloy substrates generally exhibit high brittleness and high friction coefficient which have seriously restricted their extensive applications. For the ceramic coatings are hard to bear heavy impingement and mechanical deformation due to their high brittleness. Furthermore, the high friction coefficient of ceramic coatings can easily cause the wear damage of counterpart materials. Therefore, overcoming the challenges of improving the toughness and reducing the friction coefficient of PEO ceramic coatings is of great significance which can bring about broad application prospect in tribology.

In recent years, many researchers have done a lot of work to reduce the friction coefficient of PEO ceramic coatings. The successful methods include employing liquid lubricants, introducing post treatment such as spraying, vacuum impregnation, PVD and CVD to form a self-lubricating duplex coating and one-step preparation of self-lubricating composite coating. However, none of the methods is so perfect in applications.

Employing liquid lubricants can decrease the friction coefficient of PEO coatings and consequently reduce the wear damage of counterpart materials. The wear resistance and wear life of the PEO coatings can also be enhanced in liquid lubricated conditions, ascribed to the anti-wear ceramic coatings and friction-reducing liquid lubricants. However, employing liquid lubricants may contaminate the workpieces, especially in precise instruments. Furthermore, in rigid and severe working conditions, such as high vacuum, high temperature, chemical and radioactive environments, liquid lubricants often do not function.

The duplex coatings produced by multi-step preparations can not only decrease the friction coefficient and wear rate sharply, but also avoid the shortcomings of liquid lubricants. However, these coatings are generally comprised of two layers: an inner PEO ceramic layer and an outer solid lubricant layer. When the outer layer is worn through, the tribological properties will be back to its original level. Moreover, the coating processes are too complicated and generally employ high temperature which may degrade the coatings and/or substrates.

As for the composite coatings, the solid lubricant micro- and nanoparticles are embedded in the ceramic coatings and play a role as friction-reducing agent during the whole sliding time before complete removal of the coatings. The coating process is simple and convenient, but according to recent studies, the thicknesses of yielded composite coatings are only in the range of 13-23 μm, which are far less than that of the original PEO coatings. As a result, the load bearing capacity and wear life of the composite coatings will be limited due to the low coating thickness.

Therefore, a high quality coating is still worth investigating, which has a lower friction coefficient and wear rate, a longer wear life, good mechanical properties and with a simple, cheap, effective fabricating method. More improvements should be done for the PEO technique which is of great significance in different tribological applications.

Acknowledgements

The authors would like to acknowledge the "Hundred Talents Program" of Chinese Academy of Sciences (J. Liang) and National Natural Science Foundation of China (No. 51241006) for funding this work.

Author details

Qingbiao Li[1,2], Jun Liang[1*] and Qing Wang[2]

*Address all correspondence to: jliang@licp.cas.cn

1 State Key Laboratory of Solid Lubrication, Lanzhou Institute of Chemical Physics, Chinese Academy of Sciences, Lanzhou, PR China

2 School of Science, Lanzhou University of Technology, Lanzhou, PR China

References

[1] Helmersson U, Lattemann M, Bohlmark J, Ehiasarian AP, Gudmundsson JT. Ionized physical vapor deposition (IPVD): A review of technology and applications, Thin Solid Films 2006; 513: 1–24.

[2] Wang LS, Zhang XZ, Zhao SQ, Zhou GY, Zhou YL, Qi JJ. Synthesis of well-aligned ZnO nanowires by simple physical vapor deposition on c-oriented ZnO thin films without catalysts or additives, Applied Physics Letters 2005; 86: 241081–241083.

[3] Choy KL. Chemical vapour deposition of coatings, Progress in Materials Science 2003; 48: 57–170.

[4] Emmerich R, Enders B, Martin H, Stippich F, Wolf GK, Andersen PE, Kúdelha J, Lukác P, Hasuyama H, Shima Y. Corrosion protection ability of Al2O3 coatings deposit-

ed with ion beam assisted deposition, Surface and Coatings Technology 1997; 89: 47-51.

[5] Singh L, Chawla V, Grewal JS. A Review on Detonation Gun Sprayed Coatings, Journal of Minerals & Materials Characterization & Engineering 2012; 11(3): 243–265.

[6] Markov GA, Mironva MK, Potapova OG, Lzvetiya A, Nauk K. Neogranicheskie Materialy, 1983; 19(7): 110–118.

[7] Rudnev VS, Yarvaya TP, Morozova VP, Rudnev AS, Gordienko PS. Microplasma oxidation of aluminum alloy in aqueous electrolytes with polyphosphate-Mg2+ complex anions, Journal of Protection of Metals 1999; 35(5): 473–479.

[8] Van TB, Brown SD, Wirtz GP. Anode spark reaction products in aluminate, tungstate and silicate, Journal of Bulletins American Ceramic Society 1977; 56(6): 563–566.

[9] Voevodin AA, Yerokhin AL, Lyubimov VV. Characterization of wear protective Al-Si-O coatings formed on Al-based alloys by micro-arc discharge treatment, Surface and Coatings Technology 1996; 86-87: 516–521.

[10] Wernick S, Pinner R, Sheasby PG. The surface treatment and finishing of aluminium and its alloys, 5th edn; 1987, Teddington, Finishing Publications Ltd.

[11] Henley VF. Anodic oxidation of aluminium and its alloys, 1982, Oxford, Pergamon Press.

[12] Walsh FC, Low CTJ, Wood RJK, Stevens KT, Archer J, Poeton AR, Ryder A. Plasma electrolytic oxidation (PEO) for production of anodised coatings on lightweight metal (Al, Mg, Ti) alloys, Transactions of the Institute of Metal Finishing 2009; 87: 122–135.

[13] Guntershulze A, Betz H. Elektroliticheskie Kondensatory (Electrolytic Capacitors), Moscow: Obornizidat, 1938.

[14] Li JZ, Shao ZC, Tian YW, Kang FD, Zhai YC. Application of microarc oxidation for Al、Mg, Ti and their alloys, Corrosion Science and Protection Technology 2004; 16(4): 218–221.

[15] Yerokhin AL, Nie X, Leyland A, Matthews A, Dowey SJ. Plasma electrolysis for surface engineering, Surface and Coatings Technology 1999; 122: 73–93.

[16] Wirtz GP, Brown SD, Kriven WM. Materials Manufacture Processes 1991; 6:87.

[17] Curran JA, Clyne TW. Porosity in plasma electrolytic oxide coatings, Acta Materialia 2006; 54: 1985–1993.

[18] Sun XT, Jiang ZH, Xin SG, Yao ZP. Composition and mechanical properties of hard ceramic coating containing α-Al2O3 produced by microarc oxidation on Ti-6Al-4V alloy, Thin Solid Films 2005; 471: 194–199.

[19] Xin SG, Song LX, Zhao RG, Hu XF. Composition and thermal properties of the coating containing mullite and alumina, Materials Chemistry and Physics 2006; 97: 132–136.

[20] Liang J, Guo BG, Tian J, Liu HW, Zhou JF, Xu T. Effect of potassium fluoride in electrolytic solution on the structure and properties of microarc oxidation coatings on magnesium alloy, Applied Surface Science 2005; 252: 345–351.

[21] Liang J, Hu LT, Hao JC, Preparation and characterization of oxide films containing crystalline TiO2 on magnesium alloy by plasma electrolytic oxidation, Electrochimica Acta 2007; 52: 4836–4840.

[22] Shin KR, Ko YG, Shin DH. Effect of electrolyte on surface properties of pure titanium coated by plasma electrolytic oxidation, Journal of Alloys and Compounds 2011; 509: s478–s481.

[23] Becerik DA, Ayday A, Kumruoğlu LC, Kurnaz SC, Özel A. The Effects of Na2SiO3 Concentration on the Properties of Plasma Electrolytic Oxidation Coatings on 6060 Aluminum Alloy, JMEPEG 2012; 21:1426–1430.

[24] Forno AD, Bestetti M. Effect of the electrolytic solution composition on the performance of micro-arc anodic oxidation films formed on AM60B magnesium alloy, Surface and Coatings Technology 2010; 205: 1783–1788.

[25] Shi XL, Wang QL, Wang FS, Ge SR, Effects of electrolytic concentration on properties of micro-arc film on Ti6Al4V alloy, Mining Science and Technology 2009; 19: 220–224.

[26] Wei CB, Tian XB, Yang SQ, Wang XB, Fu RKY, Chu PK. Anode current effects in plasma electrolytic oxidation, Surface and Coatings Technology 2007; 201: 5021–5024.

[27] Hussein RO, Nie X, Northwood DO. Influence of process parameters on electrolytic plasma discharging behaviour and aluminum oxide coating microstructure, Surface and Coatings Technology 2010; 205: 1659–1667.

[28] Yang Y, Wu H. Effects of Current Frequency on the Microstructure and Wear Resistance of Ceramic Coatings Embedded with SiC Nano-particles Produced by Micro-arc Oxidation on AZ91D Magnesium Alloy, Journal of Material Science Technology 2010; 26(10): 865–871.

[29] Khan RHU, Yerokhin A, Li X, Dong H, Matthews A. Surface characterisation of DC plasma electrolytic oxidation treated 6082 aluminium alloy: Effect of current density and electrolyte concentration, Surface and Coatings Technology 2010; 205: 1679–1688.

[30] Huang P, Wang F, Xu KW, Han Y. Mechanical properties of titania prepared by plasma electrolytic oxidation at different voltages, Surface and Coatings Technology 2007; 201: 5168–5171.

[31] Liu F, Xu JL, Yu DZ, Wang FP, Zhao LC. Effects of cathodic voltages on the structure and properties of ceramic coatings formed on NiTi alloy by micro-arc oxidation, Materials Chemistry and Physics 2010; 121: 172–177.

[32] Tang YM, Zhao XH, Jiang KS, Chen J, Zuo Y. The influences of duty cycle on the bonding strength of AZ31B magnesium alloy by microarc oxidation treatment, Surface and Coatings Technology 2010; 205: 1789–1792.

[33] Habazaki H, Tsunekawa S, Tsuji E, Nakayama T. Formation and characterization of wear-resistant PEO coatings formed on β-titanium alloy at different electrolyte temperatures, Applied Surface Science 2012; 259: 711–718.

[34] Gu YH, Bandopadhyay S, Chen CF, Guo YJ, Ning CY. Effect of oxidation time on the corrosion behavior of micro-arc oxidation produced AZ31 magnesium alloys in simulated body fluid, Journal of Alloys and Compounds 2012; 543: 109–117.

[35] Wang YM, Jiang BL, Lei TQ, Guo LX. Microarc oxidation coatings formed on Ti6Al4V in Na2SiO3 system solution: Microstructure, mechanical and tribological properties, Surface and Coatings Technology 2006; 201: 82–89.

[36] Liang J, Hu LT, Hao JC, Characterization of microarc oxidation coatings formed on AM60B magnesium alloy in silicate and phosphate electrolytes, Applied Surface Science 2007; 253: 4490–4496.

[37] Srinivasan PB, Blawert C, Dietzel W. Dry sliding wear behaviour of plasma electrolytic oxidation coated AZ91 cast magnesium alloy, Wear 2009; 266: 1241–1247.

[38] Treviño M, Garza-Montes-de-Oca NF, Pérez A, Hernández-Rodríguez MAL, Juárez A, Colás R. Wear of an aluminium alloy coated by plasma electrolytic oxidation, Surface and Coatings Technology 2012; 206: 2213–2219.

[39] Chen F, Zhou H, Chen C, Xia YJ. Study on the tribological performance of ceramic coatings on titanium alloy surfaces obtained through microarc oxidation, Progress in Organic Coatings 2009; 64: 264–267.

[40] Tian J, Luo ZZ, Qi SK, Sun XJ, Structure and antiwear behavior of micro-arc oxidized coatings on aluminum alloy, Surface and Coatings Technology 2002; 154: 1–7.

[41] Wei TB, Yan FY, Tian J. Characterization and wear- and corrosion-resistance of microarc oxidation ceramic coatings on aluminum alloy, Journal of Alloys and Compounds 2005; 389: 169–176.

[42] Zhou F, Wang Y, Ding HY, Wang ML, Yu M, Dai ZD. Friction characteristic of micro-arc oxidative Al2O3 coatings sliding against Si3N4 balls in various environments, Surface and Coatings Technology 2008; 202: 3808–3814.

[43] Zhu MH, Cai ZB, Lin XZ, Zheng JF, Luo J, Zhou ZR. Fretting wear behaviors of micro-arc oxidation coating sealed by grease, Wear 2009; 267: 299–307.

[44] Wang YM, Lei TQ, Guo LX, Jiang BL. Fretting wear behaviour of microarc oxidation coatings formed on titanium alloy against steel in unlubrication and oil lubrication, Applied Surface Science 2006; 252: 8113–8120.

[45] Erdemir A. Review of engineered tribological interfaces for improved boundary lubrication, Tribology International 2005; 38: 249–256.

[46] Duan HP, Du KQ, Yan CW, Wang FH. Electrochemical corrosion behavior of composite coatings of sealed MAO film on magnesium alloy AZ91D, Electrochimica Acta 2006; 51: 2898–2908.

[47] Wang ZJ, Wu LN, Qi YL, Cai W, Jiang ZH. Self-lubricating Al2O3/PTFE composite coating formation on surface of aluminium alloy, Surface and Coatings Technology 2010; 204: 3315– 3318.

[48] Wang YM, Jiang BL, Lei TQ, Guo LX. Microarc oxidation and spraying graphite duplex coating formed on titanium alloy for antifriction purpose, Applied Surface Science 2005; 246: 214–221.

[49] Wang YM, Jiang BL, Guo LX, Lei TQ. Tribological behavior of microarc oxidation coatings formed on titanium alloys against steel in dry and solid lubrication sliding, Applied Surface Science 2006; 252: 2989–2998.

[50] Martini C, Ceschini L, Tarterini F, Paillard JM, Curran JA. PEO layers obtained from mixed aluminate–phosphate baths on Ti–6Al–4V: Dry sliding behaviour and influence of a PTFE topcoat, Wear 2010; 269: 747–756.

[51] Nie X, Leyland A, Song HW, Yerokhin AL, Dowey SJ, Matthews A. Thickness effects on the mechanical properties of micro-arc discharge oxide coatings on aluminium alloys, Surface and Coatings Technology 1999; 116–119: 1055–1060.

[52] Wänstrand O, Larsson M, Kassman-Rudolphi Å. An experimental method for evaluation of the load-carrying capacity of coated aluminium: the influence of coating stiffness, hardness and thickness, Surface and Coatings Technology 2000; 127:107–113.

[53] Awad SH, Qian HC. Deposition of duplex Al2O3/TiN coatings on aluminum alloys for tribological applications using a combined microplasma oxidation (MPO) and arc ion plating (AIP), Wear 2006; 260: 215–222.

[54] Nie X, Wilson A, Leyland A, Matthews A. Deposition of duplex Al2O3/DLC coatings on Al alloys for tribological applications using a combined micro-arc oxidation and plasma-immersion ion implantation technique, Surface and Coatings Technology 2000; 121: 506–513.

[55] Liang J, Wang P, Hu LT, Hao JC. Tribological properties of duplex MAO/DLC coatings on magnesium alloy using combined microarc oxidation and filtered cathodic arc deposition, Materials Science and Engineering A 2007; 454–455: 164–169.

[56] Arslan E, Totik Y, Demirci EE, Efeoglu I. Wear and adhesion resistance of duplex coatings deposited on Ti6Al4V alloy using MAO and CFUBMS, Surface and Coatings Technology 2013; 214: 1–7.

[57] Aliofkhazraei M, Rouhaghdam AS. Wear and coating removal mechanism of alumina/titania nanocomposite layer fabricated by plasma electrolysis, Surface and Coatings Technology 2011; 205: S57–S62.

[58] Aliofkhazraei M, Rouhaghdam AS, Shahrabi T. Abrasive wear behaviour of Si3N4/TiO2 nanocomposite coatings fabricated by plasma electrolytic oxidation, Surface and Coatings Technology 2010; 205: S41–S46.

[59] Aliofkhazraei M, Rouhaghdam AS. Fabrication of functionally gradient nanocomposite coatings by plasma electrolytic oxidation based on variable duty cycle, Applied Surface Science 2012; 258: 2093–2097.

[60] Wu XH, Qin W, Guo Y, Xie ZY. Self-lubricative coating grown by micro-plasma oxidation on aluminum alloys in the solution of aluminate–graphite, Applied Surface Science 2008; 254: 6395–6399.

[61] Mu M, Zhou XJ, Xiao Q, Liang J, Huo XD, Preparation and tribological properties of self-lubricating TiO2/graphite composite coating on Ti6Al4V alloy, Applied Surface Science 2012; 258: 8570–8576.

[62] Guo J, Wang LP, Wang SC, Liang J, Xue QJ, Yan FY. Preparation and performance of a novel multifunctional plasma electrolytic oxidation composite coating formed on magnesium alloy, Journal of Materials Science 2009; 44:1998–2006.

[63] Mu M, Liang J, Zhou Xj, Xiao Q. One-step preparation of TiO2/MoS2 composite coating on Ti6Al4V alloy by plasma electrolytic oxidation and its tribological properties, Surface and Coatings Technology 2013; 214: 124–130.

Chapter 6

Metallic and Oxide Electrodeposition

Eric M. Garcia, Vanessa F.C. Lins and Tulio Matencio

Additional information is available at the end of the chapter

1. Introduction

Electrodeposition of metals or its oxides is one of oldest themes in electrochemical science [1-6]. The first studies on this topic are dated from early nineteenth century, using galvanic cells as a power source [3-6]. Despite the antiquity, electrodeposition remains a much studied topic. Themes as supercapacitors or electrochemical cells devices have raised considerable attention [7-17]. In this case, the electrodeposition technique is of great interest due to their unique principles and flexibility in the control of the structure and morphology of the oxide electrodes.

One of the most modern applications of oxides electrodeposition is solar cells. Investigation of the development of environmentally friendly low cost solar cells with cheaper semiconductor materials is extremely important for the development of green energy technology. The electrodeposition of Cu_2O, TiO_2 and many others oxides, is a very promissory research field due the low cost and high efficiency of this electrochemical method.

Moreover, two new topics that also deserves attention is the application of electrodeposition in solid oxide fuel cells (SOFC) and the metals recycling. In this field of study the electrodeposition also contributes very significantly. For SOFC, the electrodeposition is used to formation of protective coating on electrical interconnects. By other hand, in the metal recycling, the electrodeposition is the cheaper and simple method by obtention of metallic elements.

Thus, in this chapter will be reported some theoretical aspects about electrodeposition of metals and oxides and their applications more modern and relevant. Among these applications will be treated with one application of electrodeposition in supercapacitors, solar cells, recycling of metals and electrical interconects for solid oxide fuel cells (SOFC).

1.1. Theoretical foundation of electrodeposition

In the early stages of electrodeposition, the limiting step corresponds to electrons transfer from work electrode for metallic ions in solution. In this case the relation between the current and the overpotential for electrodeposition is given by Eq. 1 [1-4]. In this equation, F is Faraday's constant, k is a constant, C is the concentration of metal ions in solution, α corresponds to a coefficient of symmetry (near 0.5), η corresponds to overpotential, R is the ideal gas constant and T the absolute temperature, in Kelvin. There is an exponential dependence between the current and applied overpotential. Obviously that, with increasing of overpotential, the ionic current that electrolyte can supply is limited by the other processes as such material transport or electrical conductivity [2]. Through Coulomb's law, we obtain the relation of thickness with the charge density (d = MMq/nFϱ). Where, MM corresponds to the molecular weight, q is the charge density, n corresponds the charge of metal ions and ϱ is the density.

$$i = -FkC\exp\left(\frac{\alpha F\eta}{RT}\right) \tag{1}$$

In electrodeposition, is very common the use of potential versus current curves called voltammetry. The figure 1-a shows a typical cyclic voltammetry for cobalt electrodeposition on a steel electrode. The cyclic voltammetry is used to identify the potential where begins the electrodeposition and the electrodissolution. In cathodic scan, for potential more negative than -0.70 V occurs the Co^{+2} reductions for metallic cobalt onto steel electrode. In anodic scan, the cobalt dissolution begins in -0.3 V. The voltammogram shown in figure 1-a represents the cobalt electrodeposition, however, many others metallic electrodeposition follow the same pattern.

In the electrodeposition of M^{n+} ions using the aqueous media (Eq. 2) is always observed the hydrogen evolution reaction (HDR) represented by equation 2. This results principally in reducing the loading efficiency of electrodeposition [1-5].

$$M^{n+}_{(aq)} + ne^- \rightarrow M_{(s)} \tag{2}$$

$$2H^+_{(aq)} + 2e^- \rightarrow H_{2(g)} \tag{3}$$

The charge efficiency is an important aspect in metal electrodeposition. In this case, the parallel reaction showed in the equations 2 has a great influence in the electrodeposition. The figure 1b shows the charge efficiency for cobalt electrodeposition in pH = 1.50 (high concentration of H^+) and pH = 6.00 (low concentration of H^+). In pH = 1.50 note that the maximum efficiency (about 68%), while in pH = 6.00 the charge efficiency value can easily reach 95%. This occurs because at high concentrations of H^+, part of charge is used for promotes the reaction shown by Equation 2. In pH =6.00 the concentration of H⁺ is very low compared with cobalt concentration, thus, the cobalt electrodeposition is the principal reaction.

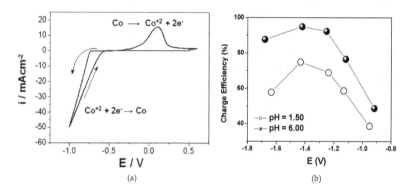

Figure 1. a) Cobalt electrodeposition on a platinum electrode. The ionic concentration of Co⁺² is 1.00 molL⁻¹ and the scan rate was 1 0mVs⁻¹. (b) Charge efficiency for cobalt electrodeposition in a range of potentials, and at pH 1.50 and 6.00.

The morphological aspects are also influenced by parallels reactions[1,2,3]. The figure 2 shows the metallic cobalt electrodeposited on ferritic steel in pH = 1.50 and 6.00. It clearly appears that the deposit at pH = 1.50 is more compact than the deposit at pH = 6.00. According to many authors this effect appears due the H⁺ reduction onto surface of growth islands. The application of the nucleation models to the initial electrodeposition stages shows that at pH=6.00, the nuclei grow progressively (progressive nucleation). SEM showed a three-dimensional nucleus growth. With the decrease in pH to 1.50, the nucleation process becomes instantaneous (instantaneous nucleation).

Figure 2. Scanning Electron Microscopy images of cobalt electrodeposited in: (a) pH= 6.00 and (b) pH 1.50. The potential applied of –1.00 V and charge density was 10.0 C cm⁻² and the electrolyte was CoSO₄ 1 molL⁻¹.

1.2. Oxide electrodeposition

In the electrodeposition of oxides, more of a method can be adopted. From a metal layer previously electrodeposited can be performed a polarization in alkaline solution such as NaOH, KOH etc. This results can be expressed by equations 4 and 5. The first step is the metal dissolution that is an electrochemical process (Eq. 4). The second step (Eq. 5) is the chemical process due precipitation of hydroxide on the surface of substrate. This method produces films with high adherence and a reduced thickness. This probably occurs due to formation of an oxide layer on a metal layer with a high surface area [4-5].

$$M_{(s)} \rightarrow M^{n+}_{(aq)} + ne^- \tag{4}$$

$$M^{n+}_{(aq)} + OH^-_{(aq)} \rightarrow M(OH)_{n(ads)} \tag{5}$$

Figure 3-a shows an electrodeposited layer of cobalt over a stainless steel. Figure 3-b represents the cobalt film after a potentiostatic polarization at 0.7 V (Ag / AgCl reference) during 200 s. The electrolyte used was NaOH 6 molL^{-1}.

The cobalt film electrodeposited showed a high contact area. This can also be visualized in the oxide/ hydroxide formed after the anodic polarization in NaOH 6 molL^{-1}.

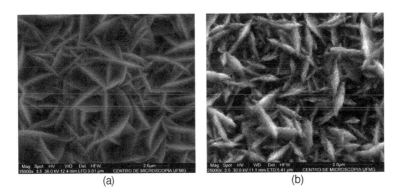

(a)   (b)

Figure 3. *Scanning electron microscopy images of (a) Cobalt electrodeposition on a steel electrode.(b) Cobalt film after a potentiostatic polarization at 0.7 V (Ag / AgCl reference) during 200 s in NaOH 6 molL^{-1}.*

Another method frequently used for electrodeposition of hydroxides on the conductive substrates is the use of nitrates as counter ions. In this case the hydroxyl ions are generated by reduction of nitrate ions in solution (Eq. 5).. The film of metallic hydroxide is then generated as shown by Eq 4.

$$NO^-_{3(aq)} + 7H_2O + 8e^- \rightarrow NH^+_{4(aq)} + 10OH^-_{(aq)} \tag{6}$$

Another method used for the formation of metal oxides is the anodic electrodeposition In this case the equation that represents this generic reaction is shows in Eq 6.

$$M_{(aq)}^{n+} + mH_2O \underset{\leftarrow}{\overset{\rightarrow}{}} Mn_2O_{n(s)} + 2mH_{(aq)}^{+} + ne^{-} \tag{7}$$

1.3. Electrodeposition morphology

Basically in the most papers, the electrodeposition can occur by passing a current fixed or by imposing a fixed potential. In both cases, the energy for the formation of nuclei for growth is given by Eq. 7 (the nucleation energy). In this equation, N^* is the number of atoms per nucleus growth (cluster), z_i is the charge of the metal ion, e_0 is the elementary charge and $|\eta|$ corresponds to the overpotential applied. The second term represents the increased surface tension caused by the addition of ad-atoms. Thus, the greater number of atoms larger the cluster size as shows in the Eq. 7. Optimizing the nucleation energy appropriately ($d\Delta G/d\eta=0$) we get the expression for maximum atoms number that each nuclei may contain (Eq. 8). where ε is the surface energy and s is the area of each atom.

$$\Delta G = -N^* z_i e_0 |\eta| + \varphi(N) \tag{8}$$

$$N_{crit} = \frac{\pi \varepsilon^2 s}{(z_i e_0 \eta)^2} \tag{9}$$

Thus, it is not difficult to observe that the higher overpotential, lower the number of atoms per growth nucleus and consequently the smaller grain size obtained. The electrodeposition created under low overpotential have smaller grains, while samples prepared under high overpotential are formed by pyramidal structures. In fact this observation can also be seen in Figure 2. This Figure shows the SEM of electrodeposited cobalt obtained in 100 and 200 mAcm^{-2}. In the higher current density, the grain size is much smaller.

Now we can see that, unlike other deposition techniques, in the electrodeposition, the characteristics of formed film are related with simple parameters as pH and overpotential applied. This makes with this method became versatile and inexpensive compared to other deposition methods.

2. Modern applications of metallic electrodeposition

With the electrodeposition is possible to achieve very thin layers of metal over the other metal or another electrical conductor material. Since that a metallic film is onto conductive substrate is possible formation of oxide layers through electrochemical methods.

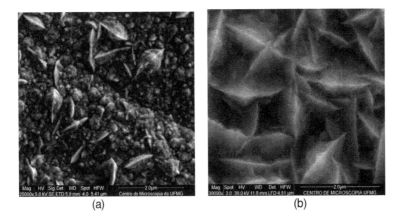

Figure 4. SEM images of cobalt electrodeposited from 0.5 molL^{-1} sulphate cobalt at 200mAcm^{-2} and (b) at 100 mAcm^{-2}.

2.1. Supercacitors

The energy always played an important role in human being's life [4]. Thus, it is necessary to study about renewable energy sources to reduce the energy consumption. Because this, the capacitors are used in the transport systems as a mean to store energy and reuse it during short periodic intervals. Basically, the conventional capacitors consist of two conducting electrodes separated by an insulating dielectric material and The application of this electrical device is the storage of energy.

When a voltage is applied to a capacitor, opposite charges accumulate on the surfaces of each electrode. The differential capacitance is defined as C = dQ/dV where V is the difference of potential between the capacitor plates and Q is the accumulated charge in the active surface of capacitor material. The geometric relation of capacitance can be providing by equation 3 [6-7]. Where A and d are the geometric area and distance between capacitors plates respectively. The ε and ε° are the dielectric constant between the plates (no unit) and the vacuum permittivity (8.8 x 10^{-12} Fm^{-1}) respectively.

$$C = \frac{A\varepsilon\varepsilon^0}{d} \qquad (10)$$

In this form C is given in Faraday (F). Thus, for practical applications the capacitors plates must have high area and very small separation distance. The first capacitors of carbon materials can provide the specific capacitances of 15 – 50 μFcm^{-2} [8]. The materials with capacitance densities of one, two or more hundreds of Fcm^{-2} or Fg^{-1} have been denominated as"supercapacitors" or "ultracapacitors"[4-8]. There are two principal types of supercapacitors: (a) the double-layer capacitor, and (b) the redox pseudocapacitor, the latter being developed in this paper. The high performances of capacitors that work with reversible redox reaction are known as supercapacitors. The most used and commercial supercapacitors are basically made of an oxide semiconductor with a reversible redox couple. This oxide has to show a transition metal ion, which must be able to assume different valences and strong bonding power.

Among the oxide materials for application in supercapacitors, ruthenium and iridium oxides have achieved more attention. In many cases the capacitance depends on the preparation method and the deposition of oxides materials. Capacitances up to 500 F/g [5] or 720 F/g are reported for amorphous water-containing ruthenium oxides [6-9]. The great disadvantages of RuO_2 is the high cost and its low porosity structure [9-10]. Thus in recent works great efforts were undertaken in order to find new and cheaper materials [4-6]. A cheaper and widely used alternative is the activated carbon. The specific capacitance of this material is about 800 F/g [6]. In this context the researchers have tried to develop a material having a low cost, high surface area, high conductivity and high stability in alkaline medium.

2.1.1. Cobalt oxides

Cobalt oxides are attractive in view of their layered structure and their reversible electrochemical reactions. The possibility of enhanced performance through different preparative methods, also interests the scientists. The spinel cobalt oxide Co_3O_4, for example, can be obtained from hydroxide cobalt (II) previously deposited onto a conductive substrate (steel for example). Is enough a thermal oxidation in 400 degrees in air atmosphere, for formation of Co_3O_4. In the Co_3O_4 oxide, the reversibility of both redox process (Co^{+2}/Co^{+3}) (Eq. 4) and (Eq. 5) (Co^{+3}/Co^{+4}) in KOH 6 molL^{-1} is very high and promising for capacitive applications in electrochemical devices [7-14].

$$Co^{+2}_{(net)} + OH^{-}_{(aq)} \underset{\leftarrow}{\overset{\rightarrow}{}} [Co-OH_{(ads)}]^{+3} + e^{-} \tag{11}$$

$$[Co-OH_{(ads)}]^{+3}_{(ads)} + OH^{-}_{(aq)} \underset{\leftarrow}{\overset{\rightarrow}{}} [Co-O_{(ads)}]^{+4} + H_2O + e^{-} \tag{12}$$

In figure 3 is shown a schematic of obtaining Co_3O_4. The previously electrodeposited cobalt can be subjected to a anodic polarization in solution of KOH 6molL^{-1}. With a thermal treatment of 20 hours at a temperature of 400 °C the phase Co_3O_4 is formed onto substrate. This procedure leads to the formation of a layer of chemical composition of $Co(OH)_2$ [8-9]. The Figure 5 represents a cyclic voltammetry of a steel electrode coated with metallic cobalt in KOH 6molL^{-1}. Note that the first peak (around -0.6 V) is related with the electrodissolution of cobalt (Eq. 13) that leads to the formation of $Co(OH)_2$ (Eq. 14) [11-12].

$$Co_{(s)} \rightarrow Co^{2+}_{(aq)} + 2e^{-} \tag{13}$$

$$Co^{2+}_{(aq)} + OH^{-}_{(aq)} \rightarrow Co(OH)_{2(ads)} \tag{14}$$

The second peak, around 0.0 V, is related to the formation of $Co(OH)_3$ (Eq.15) [6-10]. The scheme showed in figure 5-b shows the obtention of Co_3O_4 from thermal oxidation.

$$Co(OH)_{2(ads)} + OH^{-}_{(aq)} \rightarrow Co(OH)_{3(ads)} + 1e^{-} \tag{15}$$

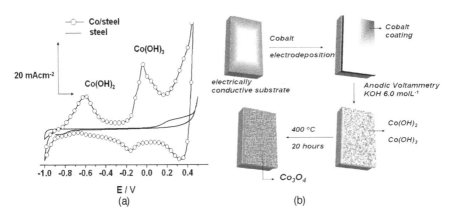

Figure 5. a) A cyclic voltammetry of a steel electrode coated with metallic cobalt in KOH 6molL⁻¹. (b) The obtention of Co_3O_4 from thermal oxidation.

There is also a method for the electrodeposition of a layer of $Co(OH)_2$ without the need for an alkaline solution. The $Co(OH)_2$ can be electrodeposited from a solution of cobalt nitrate. The many papers described this electrodeposition. The first step is the reduction of nitrate ions (Eq. 5). The reaction shown in the equation 5 promotes the alkalinization of electrical interface of metal "M". In a chemical step the Co(OH) is formed onto "M"(Eq. 6) [13-14].

$$NO^{-}_{3(aq)} + H_2O + 2e^{-} \rightarrow NO_{2(g)} + 2OH^{-}_{(aq)} \tag{16}$$

$$Co^{+2}_{(aq)} + 2OH^{-}_{(aq)} \xrightarrow{M} M / Co(OH)_{(s)} \tag{17}$$

In the vast majority of papers about supercapacitors based on cobalt oxides at least one of the steps corresponds to electrodeposition. In an interesting case in the literature, Kung et al [13] performed the electrodeposition of $Co(OH)_2$ on a Ti plate as shown by Eq.6. To obtain the Co_3O_4, the film of $Co(OH)_2$ it was submitted to an atmosphere ozone. They obtained an excellent capacitance specific value around 1000Fg⁻¹. Practically the same method however with another the substrate, Zhou et al [14] obtained a specific capacitance of 2700Fg⁻¹. In this case the substrate was Ni. Are shown in table 1 some values of specific capacitance for oxides and hydroxides of cobalt found most frequently in literature.

Method for Co(OH)₂	Substrate	Specific capacitance	Solution	Method for Co₃O₄	Reference
Potentiostatic	Ti plates	1000 F g⁻¹	Co(NO₃)₂	Ozone	[13]
Potentiostatic	Ni sheets	2700 F g⁻¹	Co(NO₃)₂	-	[14]
Cyclic voltammetry	Co electrodeposit	310 Fg⁻¹	LiOH	150 °C	[15]

Table 1. Some values of specific capacitance for oxides and hydroxides of cobalt obtained by electrodeposition or electrodeposition/calcination.

2.1.2. Manganese oxides

One promising material is hydrated amorphous or nanocrystalline manganese oxide, MnO_2 nH_2O this material has exhibited capacitances exceeding 200 F/g in solutions of several alkali salts, such as LiCl, NaCl, and KCl [16]. Chang and Tsai [17] have reported supercapacitance of 240 Fg⁻¹ for hydrous MnO_2 synthesized by potentiostatic method. By other hand Feng et al [16] reported supercapacitors of 521 Fg⁻¹ for MnO_2 multilayer nanosheets prepared galvanostatically. Both in galvanostatic or potentiostatic electrodeposition, the anodic electrodeposition for formation of MnO_2 can be express by Eq. 7:

$$Mn_{(aq)}^{+2} + 2H_2O \underset{\leftarrow}{\overset{\rightarrow}{}} MnO_{2(s)} + 4H_{(aq)}^+ + 2e^- \tag{18}$$

The model more relevant proposed to describe the charge and discharge cycles of MnO_2 at electrolyte constituted by Na_2SO_4 is given in Eq. 8. This equation demonstrated that the capacitance of MnO2 is result of redox reaction that occurs in its surface. This redox reaction corresponds of electrochemical adsorption/desorption of Na⁺ ions [16-17].

$$MnO_{2(s)} + Na_{(aq)}^+ + e^- \underset{\leftarrow}{\overset{\rightarrow}{}} (MnO_2^- Na^+)_{ads} \tag{19}$$

A fact which makes the MnO_2 becomes most promising as supercapacitor, is the recycling Zn-MnO_2 batteries. Few studies in the literature report the recycling of Zn-MnO_2 batteries [18]. Thus, this line of research becomes extremely important to make supercapacitors based in MnO_2 more environmentally correct.

2.3. Solar cells

Photovoltaic (PV) cells are made up of two semi-conductor layers and one electrolyte. One layer containing a positive charge, the other a negative charge [19-22]. The n-type semicon-ductor receive radiation hv, thus valence electron is promoted to conduction band. One potential difference is established between the n-type semiconductor and the p-type semicon-ductor. This causes a photocurrent to flow through of the system. Basically, the electrolyte is used for circulation of ionic current which restores the nature of semiconductors. The figure 6 shows the scheme of photovoltaic cell. The researches are basically focuses on development of semiconductor electrodes and electrolytes cheaper and more efficient. Thus in this context the electrodeposition can contribute greatly. This is because the electrodeposition is a method simple, practical and inexpensive to produce both p-type as n-type [21,22].

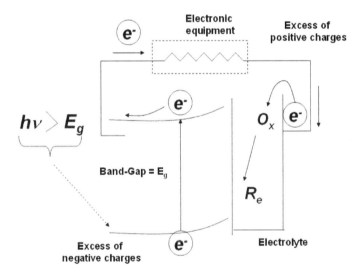

Figure 6. Representation of a photovoltaic cell using two semiconductor electrodes and an electrolyte.

2.3.1. Semiconductive Cu_2O

Among various transition metal oxides, cuprous oxide (Cu_2O) has attracted much attention. This because the Cu_2O it is a non toxic and inexpensive semiconductor material. Cu_2O is a direct band gap (1.80 eV) (Figure 7) semiconductor material and has a high absorption in the visible region of the solar spectrum [19,22,23].

The Cu_2O may be a p-type semiconductor well as the n-type. In Cu_2O-p there are vacancies of cuprous ions (Cu^+) and in Cu_2O-n, there are oxygen vacancies. Many papers on the Cu_2O-p show that this material has a lower efficiency than 2% attributed to heterojunction represented by metallic cooper (Cu/Cu_2O-p) [22]. Thus, many studies have been focused on Cu_2O-n.

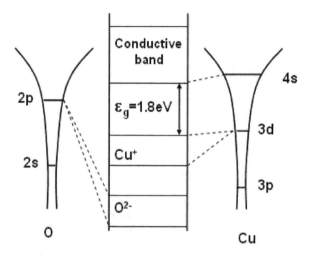

Figure 7. Representation of energy diagram for formation of Cu₂O.

Particularly, the obtention of *n*-type or *p*-type semiconductor can be controlled from pH of electrodeposition bath (Figure 8).

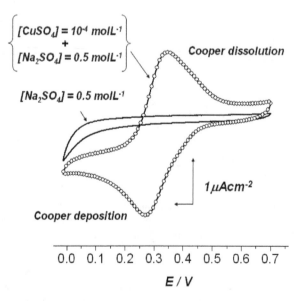

Figure 8. The cyclic voltammetry of a Pt electrode in Na₂SO₄ and CuSO₄ + Na₂SO₄ solution.

In copper electrodeposition several mechanisms are proposed. Considering the direct electro-deposition, the copper II ions are electrodeposited on one step as shown in Eq. 20. Siripala et al. [22] discusses the formation of an intermediate in the electrodeposition of copper. In this case the copper electrodeposition occurs through Cu+ forming the CuOH adsorbed (Eq. 21 and 22). Chemical occurs in one step the formation of Cu_2O (Eq. 23).

$$Cu^{+2}_{(aq)} + 2e^- \rightarrow Cu_{(s)} \tag{20}$$

$$Cu^{+2}_{(aq)} + 1e^- \rightarrow Cu^+_{(aq)} \tag{21}$$

$$Cu^+_{(aq)} + OH^-_{(aq)} \rightarrow CuOH_{(ads)} \tag{22}$$

$$CuOH_{(ads)} + CuOH_{(ads)} \rightarrow Cu_2O + H_2O \tag{23}$$

The reason for formation of Cu_2O is due to oxygen reduction forming peroxide. Among other electrochemical steps, one of the possible reactions in this case is shows in Eq. 24. The presence of Cu^+ certainly favor the formation of Cu_2O.

$$H_2O_{2(aq)} + Cu^{+2}_{(aq)} \rightarrow 2Cu^+_{(aq)} + 2H^+_{(aq)} + O_{2(g)} \tag{24}$$

According to Siripala et al. the type semiconductor (p or n) can be determined by ph of solution. If the copper electrodeposition occurs in pH below of 6.00 is formed preferably the Cu_2O-n. For pH above 6.00, electrodeposition of Cu_2O-p occurs preferably [22,23].

2.3.1. Semiconductive TiO_2

Many papers have discussed the dye-sensitized solar cell (DSC) due its promising efficiency in very low cost [19-25]. In the literature is found cathodic electrodeposition of TiO_2 nanopar-ticles on the optically transparent fluorine doped tin oxide-coated (FTO) glass [25]. This type of the oxide electrodeposition is already prepared. The deposition is made due to current of electrophoresis. In electrodeposition in pH higher than the zero charge point (PCZ) the particles TiO_2 are negatively charged. Thus one cathodic electrodeposition is unfavorable due to electrostatic repulsion. In low pH there is a tendency for formation of bubbles of hydrogen gas due to reduction of H + ions. This causes a partial occupation of the electrode surface by decreasing the efficiency of the process. The electrodeposition of TiO_2 on the FTO is pictured in the figure 9.

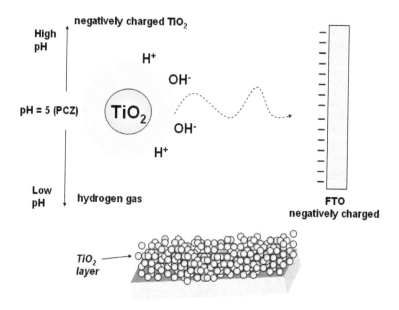

Figure 9. Representation of TiO$_2$ particles electrodeposition on F TO.

The-Vinh et al. [25] showed that the TiO$_2$/SiO$_2$ nanocomposite electrode facilitated the increase of ca. 20% of photocurrent density and ca. 30% of photovoltaic efficiency in comparison with the bare TiO$_2$ electrode. In accord to authors this electrodeposition method can be applied as an advanced microscopic way to control the electrochemical properties of the electrodes.

2.4. Electrical interconnects of Solid Oxide Fuel Cells (SOFCs)

In the current scenario where the researches are focused on cleaner energy sources and efficient, the fuel cells are gaining more space. Among the most promising fuel cells for generation of large quantity of power, are the solid oxide fuel cells (SOFCs) [26-29]. Solid oxide fuel cells (SOFCs) are solid-state devices that produce electricity by electrochemically combining fuel and air across an ionically conducting electrolyte. During operation of solid oxide fuel cell (SOFC) the anode is fed of hydrogen and cathode is supply with O$_2$ or air. At the operating temperature (600 to 800 °C), hydrogen is oxidized at the anode to H$^+$ ions. The freed electrons are conducted to the external circuit to the cathode, enabling the reduction of the oxygen ions O^{-2}(Eq.1.2) which are transported from the cathode to the anode through the electrolyte (ionic conductor). At the interface anode / electrolyte anions O^{-2} react with H$^+$ to form H$_2$O (1.3). Figure 10 depicts the scheme of operation of a SOFC.

In order to obtain high voltage and power density, a number of individual cells consisting of a porous anode, a dense thin-film electrolyte, and a porous cathode are electrically connected

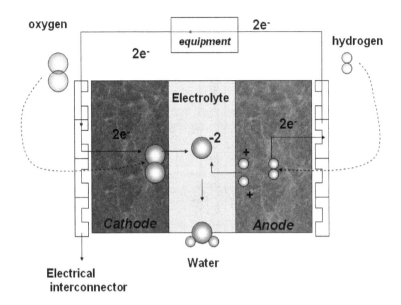

Figure 10. Simple scheme of operation and gases flow of a SOFC.

by interconnects to form a "stack". These interconnects are in contact with both electrodes (cathode and anode) and must meet a number of requirements [27-29] :

- Low area specific resistance (ASR). An acceptable value, after 40,000 working hours, is 0.10 Ohm cm^{-2}.

- Chemical stability in both atmospheres (reducing and oxidant) at high temperatures (between 600 and 1000 ºC).

- Impermeability to O_2 and H_2.

- Linear thermal expansion coefficient, LTEC, compatible with the other components of the cell (value close to 12.5 x 10^{-6} K^{-1}).

Metallic interconnects have attracted a great attention due to their high electronic and thermal conductivity and a low cost and good manufacturability compared to traditional ceramic interconnects [30]. In recent years, many works have been focused on ferritic stainless steel due to its low cost and adequate linear thermal expansion coefficient (11-12 x 10^{-6} K^{-1}) [28-30]. However, under the cathode working conditions (typically 800 ºC in air) CrO_3 evaporate from the Cr_2O_3 oxide film (Eq. 25) causing severe cell degradation [28-31].

$$\frac{1}{2}Cr_2O_{3(s)} + \frac{3}{4}O_{2(g)} \rightarrow CrO_{3(g)} \tag{25}$$

To improve the surface electrical properties and reduces the amount of chromium in the oxide film a coating of the stainless steel with semiconductor oxides has been proposed. Cobalt oxide Co_3O_4 is a promising candidate because of its interesting conductivity of 6.70 Scm^{-1} and an adequate linear thermal expansion coefficient [32-33]. A good strategy to obtain the Co_3O_4 layer over stainless steel is cobalt electrodeposition with subsequent oxidation in air at high temperatures (SOFC cathode conditions) (Figure 11). The cobalt electrodeposition can be a low cost technique. This methodology initially a cobalt layer is electroplated on steel. Under the conditions cathode (air and ~ 800) occurs the formation of Co_3O_4. In the figure 12, is shown the MEV of a cobalt layer electrodeposited on a stainless steel plate.

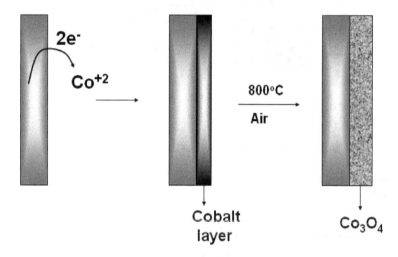

Figure 11. Scheme of Co_3O_4 obtention over stainless steel using the electrodeposition method.

The figure 13 shows that the steel without coating of cobalt subjected a strongly oxidizing conditions of cathode forms an oxide film very irregular. Also is observed that the steel surface shows the presence of chromium. Moreover the sample of coating steel with cobalt shows up very regular and without preença chromium in its surface.

In accord to Deng et al. [33] the stainless steel with electrodeposited cobalt tends to improve in high temperature and to reduce the chromium evaporation. The cobalt was oxidized to Co_3O_4. The coating not only maintains electrical contact, but it offers oxidation protection in ferritic stainless steels at lower chromium content and it is capable of significantly retarding chromium evaporation which reduces chromium poisoning of fuel cell. in accord to Wu et al [32], a uniform and smooth Mn–Co alloy can be successfully deposited on stainless steel substrate by electrodeposition. Compounds as $Mn_{1.5}Co_{1.5}O_4$ can reach an electrical conductivity of 90 Scm^{-1} to 800 degrees [33]. This shows that the electrodeposition is an excellent alternative for achieving highly conductive films at high temperatures.

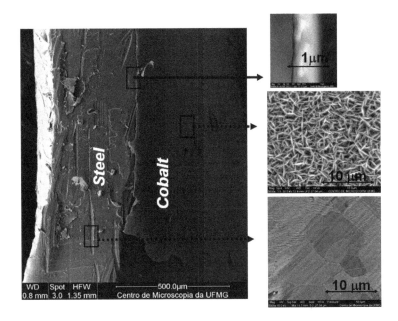

Figure 12. Scanning electron microscopy images of cobalt electrodeposited on stainless steel.

2.5. Ion-Li batteries recycling

The best way to achieve sustainable development is through recycling materials more precisely batteries. In this context, the electrodeposition takes shape more environmentally friendly. It is very interesting to note that metallic electrodeposition corresponds to a very interesting route for metal recycling. The principal metallic sources for recycling can be electronics circuits or spent batteries [6,34,35,36,37]. The Li-ion battery is a most attractive energy source for portable electronic products, such as cellular phones and laptop computers. Spinel structure $LiCoO_2$ is most used as the cathode material for Li-ion batteries due to its good performance in terms of high specific energy density and durability [1]. Thus, the Li-ion batteries are a valuable source of cobalt. The global reaction for charge and discharge of a li-ion battery with $LiCoO_2$ cathode can be represented b y Eq. 26. The $LiCoO_2$ is initially over the Al current collector. This is represented by figure 14.

$$LiCoO_2 + 6C \underset{\leftarrow}{\overset{\rightarrow}{}} Li_{(1-x)}CoO + C_6Li_x \tag{26}$$

The first step in obtaining of cobalt from Li-ion batteries is removal of $LiCoO_2$ of current collector. This is accomplished by performing a heat treatment of cathode at 400 degrees for approximately 24 hours. The powder of $LiCoO_2$ obtained is dissolved in acidic solution under constant magnetic agitation at 80 ∘C for 2 h. The cathode dissolution efficiency increases with

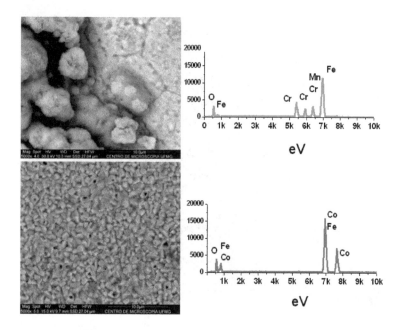

Figure 13. Scanning electron microscopy images and dispersive energy of X-ray for sample for steel coated with cobalt (low) and without cobalt (up) after 1000 hours at 850 degrees.

the increase of the acid concentration and temperature [6,34]. The addition of H_2O_2 is necessary to increase the efficiency of cathode dissolution. H_2O_2 reduces cobalt from oxidation state +III, insoluble in aqueous system, to +II, soluble in aqueous system. Considering that the active

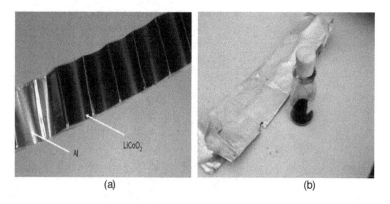

Figure 14. Photograph of the cathode of Li-ion batteries.

material is $LiCoO_2$, the cathode dissolution reaction is represented by Eq. (2) [6]. Figure 15 shows a simplified diagram of li-ion batteries recycling using the electrodeposition.

$$LiCoO_{2(s)} + \frac{1}{2}H_2O_{2(l)} + 3HCl_{(aq)} \rightarrow CoCl_{2(aq)} + \frac{1}{2}O_{2(g)} + LiCl_{(aq)} + 2H_2O_{(l)} \qquad (27)$$

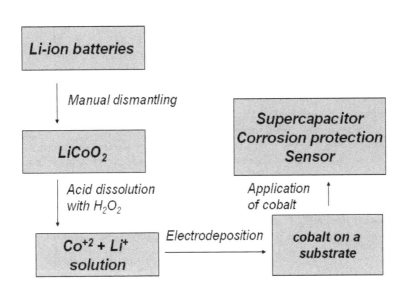

Figure 15. Simplified diagram of li-ion batteries recycling using the electrodeposition.

Although many groups working with Li-ion batteries recycling, the group most advanced in recycling via electrodeposition seems to be the group's researcher Eric M. Garcia. Garcia and other researchers perform a study of cathode Li-ion batteries recycling using the electrodeposition technique. Among several conclusions was shown that the largest charge efficiency found was 96.90% at pH 5.40. Furthermore, this research group conducted a detailed study of the mechanism of electrodeposition using electrochemistry quartz crystal microbalance technique (EQCM) [34]. In this case, it was assessed that at pH below 5, the electrodeposition of cobalt follows the direct mechanism (Eq. 13). For pH less than 2.70, cobalt electrodeposition occurs simultaneously with the reduction of protons to hydrogen [34]. In other work of Garcia and colleagues of research, it was explained the morphology of material electrodeposited with relation to pH. Although other research papers also focused on the cobalt electrodeposition as way of recycling battery Li-ion, Garcia's group pioneered the application of recycled cobalt. Garcia and other researchers associated the recycling of Li-ion battery to production of supercapacitor based on composite formed by cobalt oxides and hydroxides. The specific capacitances calculated from cyclic voltammetry and electrochemical impedance spectroscopy show a good agreement with the value of 625 Fg^{-1} [36]. Moreover Garcia et al. also proposed the application of recycled cobalt in interconnects for SOFC [37]. In this work the metallic cobalt was electrodeposited on 430 steel in order to obtain a low electrical resistance film made to Co_3O_4. After oxidation at 850 °C for 1000 h in air, the cobalt layer was transformed into the Co_3O_4 phase. On the other hand, a sample without cobalt showed the usual Cr_2O_3 and $FeCr_2O_4$ phases.

3. Conclusions

The electrodeposition remains a very important topic for technology development. Through changes in operating parameters, one can obtain metallic or oxide films with different characteristics. The electrodeposited Co_3O_4, have an excellent supercapacitive behavior with specific capacitive value around 2700 Fg^{-1}. The MnO_2 synthesized by electrodeposition method also have very good values of supercapacitance varying between 240 and 521 Fg^{-1}. In solar cells, the electrodeposition is a very promising method for electrodes fabrication. This is because the electrodeposition is a method simple, practical and inexpensive to produce both p-type as n-type. Moreover, with the advancement of Solid Oxide Fuel Cells development, the electrodeposition is once again an important method to be considered. In this case the electrodeposition is an excellent method for improved of electrical interconnects. Finally, there is also an environmental aspect of electrodeposition. This because the metals recycling of metals presents in spent batteries is made principally by electrodeposition method.

Author details

Eric M. Garcia[1], Vanessa F.C. Lins[2] and Tulio Matencio[2]

1 Federal University of São João Del Rei- unit of Sete Lagoas, Minas Gerais, Sete Lagoas, Brazil

2 Federal University of Minas Gerais, Minas Gerais, Belo Horizonte, Brazil

References

[1] Conway, B. E.Theory and Principles of Electrode Processes. Ronald, New York: 1965.

[2] Conway, B. E. Electrochemical Data. Elsevier, Amsterdam: 1952.

[3] Parsons, R. Handbook of Electrochemical Data. Butterworths, London: 1959.

[4] B J. O'M. Bockris and A. K. N. Reddy. Modern Electrochemistry. Vol. 2, Plenum, New York:1970

[5] M.B.J.G. Freitas, E.M. Garcia. Electrochemical recycling of cobalt from cathodes of spent lithium-ion batteries. Journal of Power Sources 2007; 171 953–959.

[6] Q.I.N. Chuan-li, L.U. Xing, Y.I.N. Ge-ping, B.A.I. Xu-duo, J.I.N. Zheng. Activated nitrogen-enriched carbon/carbon aerogel nanocomposites for supercapacitor applications. Transactions of Nonferrous Metals Society of China 2009;19 738-742.

[7] Q. Wang, Q. Cao, X. Wang, B. Jing, H. Kuang, L. Zhou. A high-capacity carbon prepared from renewable chicken feather biopolymer for supercapacitors. Journal of Power Sources 2013;225 101-107.

[8] Y. Zhang, J. Li, F. Kang, F. Gao, X. Wang. Fabrication and electrochemical characterization of two-dimensional ordered nanoporous manganese oxide for supercapacitor applications. International Journal of Hydrogen Energy 2012; 37 860-866.

[9] T. Yousefi, A. N. Golikand, M. H. Mashhadizadeh, M. Aghazadeh, Template-free synthesis of MnO_2 nanowires with secondary flower like structure: Characterization and supercapacitor behavior studies. Current Applied Physics 2012:12 193-198.

[10] J.B. Wu,Y. Lin, X.H. Xia, J.Y. Xu, Q.Y. Shi, Pseudocapacitive properties of electrodeposited porous nanowall Co_3O_4 film Electrochimica 2011; 56 7163– 7170.

[11] G.X. Pan, X. Xia, F. Cao, P.S. Tang, H.F. Chen. Porous Co(OH)2/Ni composite nanoflake array for high performance supercapacitors. Electrochimica Acta 2012; 63 335–340.

[12] C.W Kung, H.W. Chen, C.Y. Lin, R. Vittal, K.C. Ho. Synthesis of Co_3O_4 nanosheets via electrodeposition followed by ozone treatment and their application to high-performance supercapacitors. Journal of Power Sources 2012: 214 91-99.

[13] W.J. Zhou, M.W. Xu, D.-D. Zhao, C.L. Xu, H.L. Li. Electrodeposition and characterization of ordered mesoporous cobalt hydroxide films on different substrates for supercapacitors. Microporous and Mesoporous Materials 2009; 117 55–60.

[14] Y. Asano, T. Komatsu, K. Murashiro, K. Hoshino. Capacitance studies of cobalt compound nanowires prepared via electrodeposition. Journal of Power Sources 2011; 196 5215-5222.

[15] Z. P. Feng, G. R. Li, J.H. Zhong, Z.L. Wang, Y.N. Ou, Y.X. Tong. MnO_2 multilayer nanosheet clusters evolved from monolayer nanosheets and their predominant electrochemical properties. Electrochemistry Communications 2009; 11 706–710.

[16] S. Chou, F. Cheng, J. Chen, Electrodeposition synthesis and electrochemical properties of nanostructured δ-MnO2 films. Journal of Power Sources 2006; 162 727-734.

[17] J. Nan, D. Han, M. Cui, M. Yang, L. Pan. Recycling spent zinc manganese dioxide batteries through synthesizing Zn–Mn ferrite magnetic materials. Journal of Hazardous Materials 2006; 133 257-261.

[18] Y. Gu, X. Su, Y. Du, C. Wang. Preparation of flower-like Cu_2O nanoparticles by pulse electrodeposition and their electrocatalytic application Applied Surface Science 2010; 256 5862–5866.

[19] M. Fahoumea, O. Maghfoula, M. Aggoura, B. Hartitib, F. Chraı°bic, A. Ennaouic. Growth and characterization of ZnO thin films prepared by electrodeposition technique. Solar Energy Materials & Solar Cells 2006; 90 1437–1444.

[20] Y. Lai, Z. Chena, C. Hana, L. Jianga, F. Liua, J. Li, Y. Liu. Preparation and characterization of Sb_2Se_3 thin films by electrodeposition and annealing treatment Applied Surface Science 2012; 261 510– 514.

[21] W. Siripala. K.M.D.C. Jayathileka, J.K.D.S. Jayanetti. Low Cost Solar Cells with Electrodeposited Cuprous Oxide Journal of Bionanoscience 2009; 3 118–123.

[22] X. Han, K. Han, M. Tao, Characterization of Cl-doped n-type Cu2O prepared by electrodeposition Thin Solid Films 2010; 518 5363–5367.

[23] S. Thanikaikarasana, K. Sundarama, T. Mahalingama, S. Velumanib, J.K. Rheec. Electrodeposition and characterization of Fe doped CdSe thin films from aqueous solution. Materials Science and Engineering B 2010; 174 242–248.

[24] T.V. Nguyen, H.C. Lee, M. A. Khan, O.B. Yang. Electrodeposition of TiO_2/SiO_2 nanocomposite for dye-sensitized solar cell. Solar Energy 2007; 81 529–534.

[25] M.R. Ardigò, A. Perron, L. Combemale, O. Heintz, G. Caboche, S. Chevalier. Interface reactivity study between $La_{0.6}Sr_{0.4}Co_{0.2}Fe_{0.8}O_{(3-x)}$ (LSCF) cathode material and metallic interconnect for fuel cell Journal of Power Sources 2011; 196 2037–2045.

[26] P.Y. Chou, C.J. Ciou, Y.C. Lee, I.M. Hung. Effect of $La_{0.1}Sr_{0.9}Co_{0.5}Mn_{0.5}O_{(3-x)}$ protective coating layer on the performance of $La_{0.6}Sr_{0.4}Co_{0.8}Fe_{0.2}O_{(3-x)}$ solid oxide fuel cell cathode Journal of Power Sources 197 (2012) 12– 19

[27] Y. Zhen, S. P. Jiang, Characterization and performance of (La,Ba)(Co,Fe)O_3 cathode for solid oxide fuel cells with iron–chromium metallic interconnect. Journal of Power Sources 2008; 180 695–703.

[28] A. N. Hansson, S. Linderoth, M. Mogensen, M. A.J. Somers. Inter-diffusion between Co_3O_4 coatings and the oxide scale on Fe-22Cr. Journal of Alloys and Compounds 2007; 433 193–201.

[29] Z. Yang, G. Xia, P. Singh, J.W. Stevenson Electrical contacts between cathodes and metallic interconnects in solid oxide fuel cells. Journal of Power Sources 2006; 155 246–252.

[30] M. R. Bateni, P. Wei, X. Deng, A. Petric. Spinel coatings for UNS 430 stainless steel interconnects. Surface & Coatings Technology 2007; 201 4677–4684.

[31] J. Wu, Y. Jiang, C. Johnson, X. Liu, DC electrodeposition of Mn–Co alloys on stainless steels

[32] for SOFC interconnect application. Journal of Power Sources 2008; 177 376–385.

[33] X. Deng, P. Wei, M. R. Bateni, A Petric, Cobalt plating of high temperature stainless steel interconnects. Journal of Power Sources 2006; 160 1225–1229.

[34] E.M. Garcia, J.S. Santos, E.C. Pereira, M.B.J.G. Freitas, *Electrodeposition of cobalt from spent Li-ion battery cathodes by the electrochemistry quartz crystal microbalance technique.*. Journal of Power Sources 185 (2008) 549–553.

[35] M. B. J. G. Freitas, E. M. Garcia, V. G. Celante *Electrochemical and structural characterization of cobalt recycled from cathodes of spent Li-ion batteries* J Appl Electrochem (2009) 39:601–607

[36] E.M. Garcia, H. A. Tarôco, T. Matencio, R. Z. Domingues, J. A. F. dos Santos, R.V. Ferreira, E. Lorençon, D. Q. Lima, M.B.J.G. de Freitas. *Electrochemical recycling of cobalt from spent cathodes of lithium-ion batteries: its application as supercapacitor* J Appl Electrochem (2012) 42:361–366.

[37] E.M. Garcia, H. A. Tarôco, T. Matencio, R.Z. Domingues, J.A.F. dos Santos,M.B.J.G. de Freitas Electrochemical recycling of cobalt from spent cathodes of lithium–ion batteries: its application as coating on SOFC interconnects J Appl Electrochem (2011) 11:1373-1379.

Chapter 7

Coating Technology of Nuclear Fuel Kernels: A Multiscale View

Malin Liu

Additional information is available at the end of the chapter

1. Introduction

The coating technology of nuclear fuel kernels (NFKs) is an important part of the nuclear safety research. Now there are many new designs of nuclear reactor which based on the coated fuel particles. Among them, high temperature gas-cooled reactor (HTGR) is one of the Gen-IV reactors and has a bright future in the electricity and hydrogen production because of its superior characteristics.

The inherent safety characteristics of HTGR have been paid more attention among many nuclear reactors in the nuclear renaissance, even more adequately after the Fukushima nuclear accident. The first security assurance is the HTGR nuclear fuel element based on the coated fuel particles, so the coating process of nuclear fuel particles is one of the most important key technologies in the research on HTGR. The tristructural-isotropic (TRISO) type coated fuel particle, which has been commonly used in the current HTGR consists of a microspheric UO_2 fuel kernel surrounded by four coated layers: a porous buffer pyrolysis carbon layer (buffer PyC), an inner dense pyrolysis carbon layer (IPyC), a silicon carbide layer (SiC) and an outer dense pyrolysis carbon layer (OPyC), as shown in Fig.1. All coating layers are prepared in the spouted fluidized bed by chemical vapour deposition (CVD) method in different research groups in Germany, USA, South Korea, Japan and China [1-5].

Now, HTR-PM (high-temperature-reactor pebble-bed module), a Chinese 2×250 MWth HTR demonstration plant, is under construction in Weihai City, Shandong Province, PRC. A pilot fuel production line will be built to fabricate 300,000 pebble fuel elements per year, and each pebble fuel element contains about 15000 coated fuel particles, so the higher requirements for mass production of coated fuel particles for fuel elements are put forward. In order to optimize and scale up the coating process of fuel kernels from the lab to the factory, the multiscale study

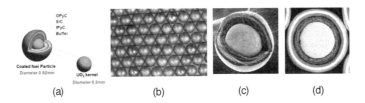

Figure 1. (a) schematic diagram of coating layers, (b) nuclear fuel kernels(~0.5mm), (c) coated fuel particle(~0.92mm), (d) Cross-section of coated fuel particle(~0.92mm)

concept of the coating process was proposed by us in the research process. In this chapter, the details of this multiscale concept will be given based on our scientific research achievements.

2. The coating methods for NFKs

The first layer deposited on the kernels is porous carbon, this is followed by a thin coating of pyrolytic carbon (a very dense form of heat-treated carbon), a layer of silicon carbide and another layer of pyrolytic carbon. Each coating process is performed at a defined temperature and for a defined duration of time. The schematic diagram of the experimental facility is shown in Fig. 2. This experimental facility can be divided into two main parts: the left part is the gas distribution system and the right part is the conical spouted bed coating system. The gas distribution system contains argon, the hydrocarbon gas (ethyne and propene), hydrogen and methyl trichlorosilane (MTS) vapor supply device, flow control system and the gas distribution device. The conical spouted bed coating system contains the spouted bed, the cooling water system, the heating system, the thermal insulation system, the gas airtight device and the temperature control device.

The spherical UO_2 kernel as fuel particles are fluidized by the fluidization gas and coated by the reactive gas in the conical spouted bed coating furnace. The thickness, coating time, coating temperature, reactive and fluidization gas of four coating layers should be determined beforehand in the coating experiments as shown in Table 1.

Coating layer	Thickness /μm	Coating time /min	Reactive+Fluidization gas	Coating temperature / °C
1. Buffer PyC	95	~2.5	Acetylene+Argon	~1260
2. Inner PyC	40	~12.5	Propylene+ Argon	~1280
3. SiC	35	~150.0	MTS+Hydrogen	~1600
4. Outer PyC	40	~12.0	Propylene+ Argon	~1300

Table 1. Experimental parameters in the coating process of TRISO particle

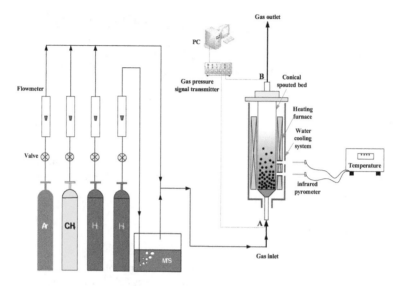

Figure 2. The schematic diagram of the spouted bed coating system

The whole coating process can be described as follows: In the pyrolytic carbon coating process, ethyne and propene as the reactive gas and argon as the fluidization gas are injected into the coating furnace, they are pyrolyzed at the high temperature about 1300ºC and then PyC layers are prepared. The SiC coated layer is prepared by MTS vapor as the reactive gas which is entrained by hydrogen and is pyrolyzed at about 1600 ºC. The integrity and property behavior of the SiC layer of the Tri-isotropic (TRISO) coated particle (CP) for high temperature reactors (HTR) are very important as the SiC layer is the main barrier for gaseous and metallic fission product release.

The fluidized bed chemical vapor deposition (FB-CVD) method is a suitable technique for preparing various kinds of films/layers on the spherical materials by initiating chemical reaction in a gas. FB-CVD has the advantage of large reactor volume to offer sufficient space with uniform mass and heat transfer condition. This technique can be used for other purposes, such as synthesizing carbon nanotube composite photocatalyst ((CNT)/Fe-Ni/TiO_2). Also, some modified method based on FB-CVD, such as plasma-enhanced FB-CVD, has been used to prepare the transparent water-repellent thin films on glass beads in modern surface engineering treatment. So the investigation of FB-CVD method is helpful and important for modern surface treatments.

3. Multiscale analysis

If we only do the coating experiments in the lab scale, it can be done in a very small reactor and the efficiency can be very high, but now the question is the how to develop coating technology from the lab to the factory? The answer is that we must consider the multiscale study of coating technology of nuclear fuel kernels. The multiscale study of TRISO-coated particle includes three scales: microscale(material microstructure scale, ~nm), mesoscale (mesoscopic scale, paticle size scale, ~mm) and macroscale(reactor size scale, ~m), as shown in Fig. 3.

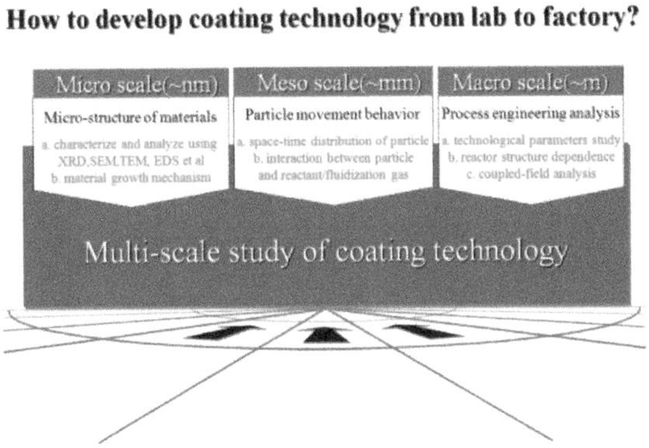

How to develop coating technology from lab to factory?

Figure 3. Multiscale study of coating techonolgy of nuclear fuel kernels

The microscale study mainly focused on the micro-structure of materials, including characterize and analyze the coating materials using XRD, SEM, TEM, EDS, et al, and then give the material growth mechanism, using it we can control the material grown as we want. This scale is the basic area of materials research. The deposition mechanism of PyC layer, the microstructure of SiC layer, and the new coating layer, such as ZrC, has been exploited in our studies.

In order to prevent the risk associated with producing particles that do not meet the requirements, a fundamental understanding of the fluidization phenomena occurring in the spouted bed coater is needed. Fluidized-bed CVD is a special technique to coat nuclear fuel particles. To ensure uniform deposits on particles, efficient contact of particles with the reactive gas must be achieved. Knowledge of the solids flow pattern in spouted bed is essential to the design of spouted bed, because the particle trajectories must meet process requirements. So the mesoscale study mainly focused on the particle movement behavior, including the space-time distribution of particles and spout-fluid bed dynamics. The domination factor is the interaction between particle and coating/fluidization gas, which influences the properties of coating materials directly.

The macroscale study mainly focused on the process engineering analysis on the whole coating system, including technological parameters study, reactor structure dependence, coupled-field analysis which include the velocity field, temperature field and material concentration field. An example is the associated research between a reactor-scale pressure changes and the coating process.

3.1. Microscale study

The microscale study mainly focused on characterizing and analyzing the micro-structure of the coating materials by using XRD, SEM, TEM, EDS, et al [6-8]. Based on the examination results, the growth mechanism of coating layers was analyzed and the relation of coating layers microstructure and the deposition technology was established. We have validated the droplet deposition mechanism in the PyC layer [9] and investigated density change reasons in the SiC layer [10]. The relationship between the temperature change and the microstructure of the SiC layer [11], and the new coating layer, such as ZrC, has been exploited.

• The coating process of PyC layer

In the PyC coating process, the ethyne and the propene are injected into a conical fluidized bed coating furnace and pyrolytic carbon is coated on the fuel particle by CVD (chemical vapor deposition). However, pyrolytic carbon has not been fully exploited in the coating process of fuel particles, so a large amount of pyrolytic carbon powder as waste will settle down in the subsequent cyclone separator. The pyrolytic carbon powder is the main solid byproduct generated in the coating process. The pyrolytic carbon powder can be collected as carbon black powder sample, which can be seen as the intermediate products and can be used to study the mechanism of the coating process of the PyC layer.

The microstructure of pyrolytic carbon powder and the PyC layer are investigated using JEM 2010 TEM (JEOL Ltd., Tokyo, Japan). The preparation method of electron microscopy sample is to suspend the PyC layer and carbon black powder in the anhydrous ethanol at first, and then disperse the sample for 15 minutes by adopting the ultrasound method.

The TEM image of the carbon black samples is showed in Fig. 4. It can be found that the pyrolytic carbon black powder is composed of the nano-spherical carbon particles. They are ring-layered nano-structured carbon particles and many layers of carbon atoms arrange tidily in the nano-spherical carbon particles.

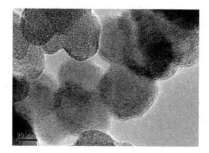

Figure 4. The TEM image of the carbon black sample

The high resolution TEM image in Fig. 5 indicate that the near-ordered structure of nano-spherical carbon particle is mainly characterized by interrupted lattice fringes of 0.34 nm in average. This value is comparable with the (002) d-spacing in graphite 2H and the (003) d-spacing in graphite 3R. But the carbon atom layer is incomplete here and the structure shows a high defect concentration. The results confirm that the carbon black sample has the layered structure and the near-ordered structure of nano-spherical carbon particle is mainly characterized by lattice fringes of 0.34 nm in average.

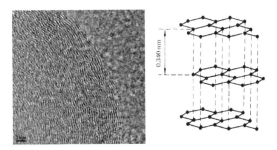

Figure 5. The high resilution TEM image of the carbon black sample

The TEM image of the deposition pyrolytic carbon samples is given in Fig.6. It can be found that agglomeration of carbon nano-particles is much greater. Numerous models have been proposed to describe the carbon deposition mechanism: molecular deposition mechanism, droplet deposition mechanism, surface decomposition theory and so on [12]. The experimental results here have shown that the gas-phase nucleation occur at first and then nano-particles deposit on the surface of fuel particle. The droplet deposition mechanism is suitable to describe the formation of the dense PyC layer in the fuel particle coating process.

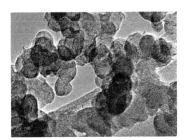

Figure 6. The TEM image of the deposition pyrolytic carbon samples

The schematic diagram of the nano-particles agglomeration is shown in Fig. 7. The nano-particles with the diameter ofabout 50 nm aggregate into the large clusters with the diameter of about 1 μm at first. Then the 1 μm clusters aggregate into the larger cluster with the diameter of 10 μm. Usually, the microstructure of the carbon black powder consists of three grades:(1) the primary particle; (2) the particle aggregates(~1μm), which is the primary structure of the carbon black; (3) the agglomerate(~10μm), which is also known as the secondary structure of carbon black. The Van der Waals force plays a major role in the agglomerating process of the nano-spherical carbon particle.

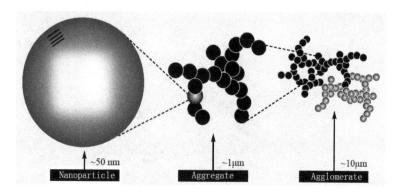

Figure 7. The schematic diagram of the nano-particles agglomeration

From the above discussion, it can be found that the droplet deposition mechanism can be used to explain the formation of the dense PyC layer in the fuel particle coating process. The nano-spherical carbon particles are generated at first, and then deposit on the surface of fuel particle.

- The coating process of SiC layer

The microstructure of the SiC coating layer are invesitgated under different coating temperatures, from 1520-1600°C. The reasons of density change are discussed based on microstructure analysis. After coating experiments, the coated particle are pressed carefully, until the SiC coated layer are crushed, then the fragments of SiC coated layer are collected carefully, as shown in Fig.8. Prior to characterization analysis, it need to heat the fragments for 10 hours at 800 °C, to remove the layers of pyrolytic carbon.

Figure 8. Appearances of SiC coating layer after crushing

The XRD analysis results of the cross-section and the surface of the SiC sample produced under different temperatures are almost the same. The peaks located at 36', 60' and 72' are the characteristic peak of the SiC, corresponding to the (111), (220) and (311) crystal face diffraction of SiC respectively. The peaks located at 41' and 75 'are corresponding to the crystal face (200), (222) diffraction of SiC respectively. The sample has a perfect crystallinity from these characterization results. The pure SiC product is obtained substantially, impurities such as carbon and silicon peaks is not obvious, so the influence of impurities such as carbon can be excluded in the possible reasons of the aforementioned density changes. Thus it can be concluded that SiC without the doped C, Si and other impurities can be prepared in the experimental MTS concentration value, in different temperature from 1 520 ℃ to 1 600 ℃.

The Raman spectra of the SiC coating layer obtained under different temperature are shown in Fig. 9. We know that the two characteristic peaks of the β-SiC are the two optical mode (peak): 796cm⁻¹, the TO(transverse optical mode) and 972cm⁻¹, LO (longitudinal optical mode) in the Brillouin zone. As can be seen from Fig.10, The Raman spectral peaks measured experimentally are the peak 1 (796cm⁻¹) and peak 2 (972cm⁻¹), which can be considered as the two characteristic peaks, therefore it can be concluded that β-SiC (3C-SiC) is obtained, and the crystal structure is the zinc-blende structure. There are no characteristic peaks of the α-SiC, it can be determined that the sample are almost pure β-SiC. So β-SiC can be generated in the experimental temperature range from 1520-1600℃.

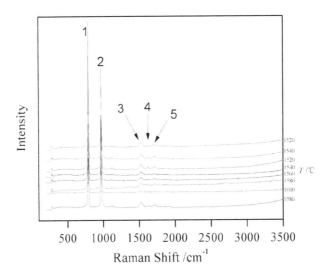

Figure 9. Raman spectrum of SiC coated layer

The secondary Raman peaks of SiC coating layer can also be obtained as: 3#(1 520 cm⁻¹), 4#(1 620 cm⁻¹), 5#(1 720 cm⁻¹) in Fig.9. It is consistant with the literature. From further analysis it can be found that the peak 5 (1 720 cm⁻¹) has a blue shift trend with the decrease in density of the sample, as shown in Fig.10. The reason can be deduced from the literature [13]. The presence of the microscopic pore structure in the sample with low density destroyed the complete lattice structure of SiC crystal, so the lattice vibration frequency increases as a result.

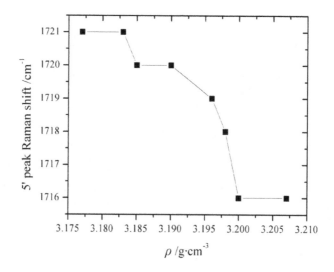

Figure 10. Secondary Raman spectrum of SiC coating layer and temperature *vs* density

The SEM image of the cross-section of the SiC coating layer with low density (ρ=3.177 g/cm³) are given and the characterization images are shown in Fig. 11. As can be seen from the figure, the coating layer of low density SiC has the the presence of a microporous structure indeed, this is consistent with the results of the Raman spectroscopy. The above results show that the porous structure should be the main reason of which the apparent density of the SiC coating layer is lower than the theoretical value.

Figure 11. SEM image of shell cross-section of SiC coating layer

The linear distribution of micropores can also be found from Fig. 11, and the linear distribution is parallel to both the inside and outside shell of the SiC coating layer approximately, i.e. the micropores are located at a same deposition surface in the growth process of the SiC coating layer. It can be assumed that during the growth of the SiC coating layer, the SiC layer grows gradually from the inner shell to the outer shell, the temperature or the fluidized state of the particles fluctuate at a certain time point, resulting of interruption occurs in the continuous SiC crystal growth, to produce a linear distribution of micropores. So these micropores are in linear distribution, and parallel to the inner and outer shell approximately. Obviously, these micropores are not conducive to the performance improvement of hindering the fission products, and should be avoided in the actual production process.

To sum up, the pure β-SiC can be prepared in the experimental temperature range from 1520-1600ºC. Microstructure analysis validates that the density change of the SiC coated layer is mainly due to the micropores in the coated layer, not the doped C, Si and other impurities. The linear distribution of the micropores is found, and all micropores are located at the same deposition surface. It can be concluded that the formation of the micropores in the coated layer is closely related to the particle fluidized state, which should be focused on in the mesoscale study.

High temperature oxidation behavior of the SiC coating layer are also investigated. High temperature oxidation experiments were carried out about 10 hours at different temperatures from 800 to 1 600 ºC at air atmosphere. The microstructure and composition were studied using different characterization methods. High-resolution SEM images are given in Fig. 12.

It showed that the oxidation began from the surface and gradually formed a punctate distribution of the oxide. The small crystal structure (~1 μm) of the SiC surface was very clear at 1000 ºC, this crystal structure was disappeared gradually at high temperatures, and oxidized to form oxides of Si in the SiC surface. There was a stacking fault stress due to the different thermal expansion coefficient and elastic modulus of SiC and Si oxides, leading to the gradual formation of cracks around the punctates. With the increasing temperature, the crack further increased the oxidation rate of SiC, finally the formed Si oxides layer fell off from the particles. It can be seen clearly that the layer was peeling when the temperature was increased to 1 400 ºC.

The SEM images of new fracture surface of SiC coating layer after oxidization at 1400 ºC are shown in Fig.13. It can seen obviously that oxidation front-end interface with many micropores are between oxidation layer with the smooth surface and the SiC coating layer with crepe-like texture. It can also be found that the oxidation layer is very thin, only about 1μm, indicating that the oxidation rate is very slow. The oxidation layer was so thin (~1μm) comparing with the original thickness(~35μm) of the SiC coating layer, so the remain SiC coating layer has the same fuction of hindering fission products in the TRISO particles.

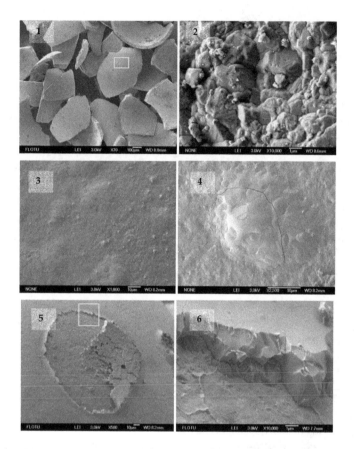

Figure 12. SEM images of SiC coating layer after oxidization at different temperature(1. 1000 °C, ×70; 2. 1000 °C, ×10000; 3. 1200 °C, ×1000; 4. 1300 °C, ×2500; 5. 1400 °C, ×500; 6. 1400 °C, ×10000)

Figure 13. SEM images of SiC fracture surface after oxidization at 1400℃

The XRD characterization results of the SiC coating material after the oxidation test of different temperatures are given in Fig. 14. The SiO_2 peaks gradually appear when the oxidization temperature is more than 1 400 °C. It showed that there was a growing SiO_2 peaks from 1 400 °C, indicating that surface oxidation was significant gradually. The relative degree of oxidation is deepened when the temperature rises. The peak is very sharp, which means that SiO_2 is already crystallized. The oxidation behavior found here is consistent with the SEM images above.

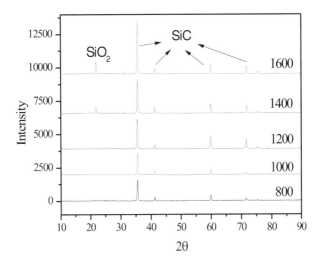

Figure 14. XRD results of the SiC coating layer after oxidization at different temperature

To sum up, the oxidization of the SiC coating layer was not obvious below 1 200 °C. The oxidization of the SiC coating layer was significant gradually from 1 400 °C, but the oxidation rate is very slow. So the SiC coating layer has good performance of hindering the fission product, even at the severe air ingress accident in the nuclear power station. The TRISO coated particles can be seen as the first security assurance of the HTGR nuclear fuel element.

3.2. Mesoscale study

The mesoscale study mainly focused on how to improve gas-solid contact efficiency, which based on a fundamental understanding of the fluidization phenomena of high density particles (UO_2, ~10.86g/cm³) occurring in the spouted bed coater. In our work, the particle movement behaviors including the space-time distribution of particles and spout-fluid bed dynamics based on Euler-Euler simulation were studied [14]. It was found that the specially designed multi-nozzle gas inlet is better than the single-nozzle gas inlet for obtaining a more uniform fluidization, which can disperse the gas to increase the gas-particle contact efficiency. Also we have established the DEM-CFD simulation platform. The spout process was simulated based on this coupling method. Some key parameters in coating process are proposed and the scale-up technology has been developed. It was emphasized that efficient contact of particles with the reactive gas must be achieved to ensure uniform deposition on particles.

• Euler-Lagrange simulation

The coating layer on the UO$_2$ fuel kernel surface is prepared by chemical vapor deposition when the gases are injected into the spouted bed coating furnace. One of the major factors to influence coating performance is the contact efficiency between the gas and the solid particles, which is determined by the hydrodynamics in the spouted bed nuclear fuel particle coated furnace. The hydrodynamics of a spouted bed are important from a CVD perspective because they directly determine the magnitude and variability of the concentration and species gradients in the zone where the reactant gases first come into contact with hot particles. It is recommended for the gas distributor to disperse the gas over the full area of the coating chamber to enhance gas-particle contact efficiency. Specifically, the particle flow in spouted bed consists of three distinct regions: a dilute phase core of upward gas-solid flow called the "spout", a surrounding region of downward quasi-static granular flow called the "annulus", and a top area called " fountain" between these aforementioned two regions as shown in Fig. 15.

Spouted bed coater

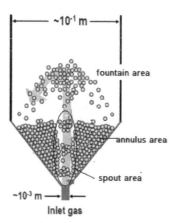

Figure 15. The spout area, annulus area and fountain area in the spouted bed coater

Generally, one particle is flowed up in spout area, and contacted with the reaction gas, coated by the CVD products, and then, falling down into the annulus area, which can not contact with the reaction gas. So the key parameters in coating process are the volume of spout area (V_s), the oscillation cycle period of particle clusters (T_c), the single particle cycle time(T_p) and the time period of particle in spout area in a particle cycle time(T_s). It is difficult to measure these key parameters in experiments, but it is easy to obtain them in the simulation results.

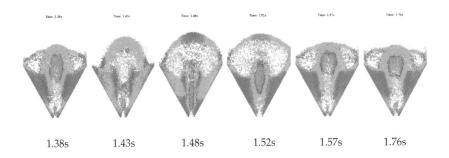

| 1.38s | 1.43s | 1.48s | 1.52s | 1.57s | 1.76s |

Figure 16. The periodical change of particle movement in the spout process(Red: high particle velocity; Blue: low particle velocity)

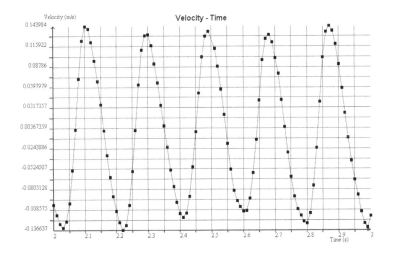

Figure 17. The periodical change of average particle vertical velocity in the spout process

The Discrete Element Method (DEM), which can resolve particle flow behaviors at an individual particle level, has been widely used for studying granular-fluid flow in fluidized-bed. The k–ε turbulence model, coupling with the DEM was used to simulate the particle spout behaviours in the coating process. The typical simualtion results in the conventional spouted bed are given in Fig.16 and Fig. 17. The simulation results indicate very rapid motion of the particles within the spout, with typical velocity of nearly 1 m/s. The oscillation cycle period of particle clusters (T_c) can also be obtained as 190ms (1.38s -1.57s is one period and 1.57s-1.76s is one period in Fig. 16). The periodical change of average particle vertical velocity in the spout process in Fig. 17 can also validate that the oscillation cycle period of particle clusters is 190ms.

The single particle cycle time(T_p) is an important factor in the coating process, which can also be obained form the simulation results of the particle movement trajectory, as shown in Fig. 18. It can be found that the cycle index is 5 from 1s to 5s, so the cycle time can be obtained as 4.0s/5=0.8s. It can also be found that the time period of particle in spout area (the movement trajectory is indicated as red color) in one particle cycle time(T_s) is 40 ms, which is very short comparing with the single particle cycle time(T_p). It means that in the whole coating process, the growth time of the coating layer occupies only 5%(=40ms/800ms) of the total time of the coated process. The conventional spouted bed should be improved to obtain a high gas-solid contact efficiency.

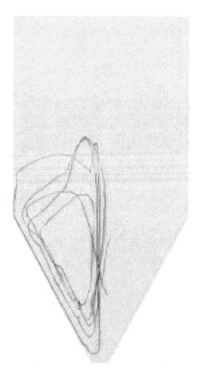

Figure 18. The particle movement trajectory in the spout process(1s-5s)

To sum up, the DEM-CFD simulation results can give the particle movement behaviors in details (T_c=190ms, T_p=800ms, T_s=40ms), so it has great potential in optimizing and scaling up the particle coating technology. It is obvious that the expansion of the spout area and reduction of the cycle time will increase the coating efficiency and improve the performance of the coating layer. These methods should be considered as the basic principles of optimization and scale up of the coating technology.

• Euler-Euler simulation

To enhance the gas solid contact efficiency, the hydrodynamics of a three-dimensional conical spouted bed was studied using an Eulerian-Eulerian two-fluid model (TFM) incorporating the kinetic theory of granular flows. Four designs with traditional single-nozzle inlet, modified single-nozzle inlet, multi-nozzle inlet and swirl flow design inlet, as shown in Fig. 19, were simulated and compared.

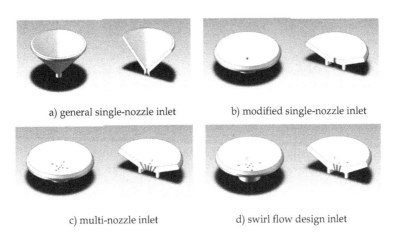

a) general single-nozzle inlet   b) modified single-nozzle inlet

c) multi-nozzle inlet   d) swirl flow design inlet

Figure 19. The pictures of different inlet designs

The typical simulation results are given in Fig. 20. The spout area is increased significantly with the multi-nozzle inlet. It can be found that the accumulation of particles near the wall disappears with a swirl flow design inlet, expecially at the bottom of the coater. These results illustrate that the design of the inlet is helpful to reduce the particle aggregation close to the wall and the spout area is enlarged as a result. The particle entrainment in the spouted bed with swirl flow design inlet increases significantly and the maximum spouted height is decreased at the same superficial gas velocity (0.6m/s), so T_s will be increased and the gas-solid contacting efficiency will be enhanced as a result.

To sum up, the swirl flow designed multi-nozzle gas inlet is better than any other gas inlets for obtaining a more uniform fluidization, which can disperse the gas to increase the gas-particle contact efficiency. In the future, the coating experimental results should be given to validate the good performance of this newly designed inlet with swirl flow multi-hole.

It should be indicated that the necessity of simulation for this process is obvious [15]. The new design and the details of particle movement can be obtained before experiments, so a lot of time and money can be saved in the research process, and the results can be used as the targeted initial conditions for further optimization in experiments. On the other hand, how to obtain accurate and meaningful simulation results is very important, it is based on the basical studies of mathematical models and numerical simulation method.

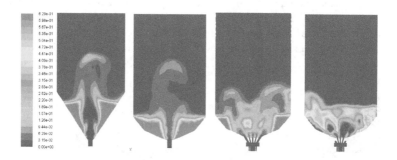

Figure 20. Profiles of the solid holdup under different inlet designs in the spouted bed at the same superficial gas velocity (0.6m/s)

3.3. Macroscale study

The macroscale study mainly focused on the process engineering analysis on the whole coating system, such as the pressure fluctuation analysis in the coating process [16, 17]. An example we have completed is the associated research between a reactor-scale pressure changes and coating process [18]. In this research, I have proposed a relationship about the change of pressure drop and the change of particle properties before and after the coating process of each layer. A convenient method for real-time monitoring the fluidized state of the particles in a high-temperature coating process is proposed based on this relationship. It will descibed in details as follows.

The conical spouted bed coating furnace is the main part of the coating system, which was designed carefully in experiments. Operations in particle spouting are sensitive to the geometry of the equipment and particle diameter/density. The following geometric parameters for stable operations in spouting bed are important, as shown in Fig. 21: (1) Cone included angle (γ), (2) Inlet diameter (D_o), (3) Column diameter (D_c), (4) Height of conical part (H_c). The pressure drop can be related to this parameters. There are also some parameters related to the particle properties, such as particle diameter d_p, static bed height H_o, particle density ρ_p, which are also very important to determinate the spouted bed hydrodynamics. The pressure drop changed with the superficial gas velocity and different stages are illustrated in Fig. 22 in details. The stable spouting state is the optimum state to achieve the efficient and uniform contact of particles with the reactive gas phase, Therefore, particles should remain the stable spouting state in the conical spouted bed coating furnace and ΔP_s is expected to be found during the whole coating process.

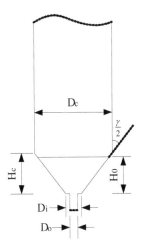

Figure 21. The geometric factors of the spouted bed using in the coating process

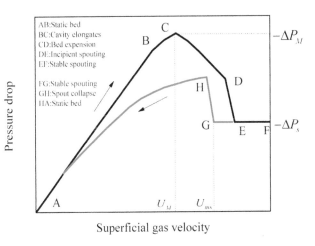

Superficial gas velocity

Figure 22. Pressure drop change in spouted bed with different superficial gas velocity

All the particles and graphite column tube are in the glowing red state at 1200-1600°C. So it is unfeasible to obtain the pressure signals in the gas solid flow in the high temperature graphite column tube directly. But it is easy to measure the pressure signals at both ends of the coating furnace in the real coating process, for example points A&B are selected for measuring the pressure signals in our experiments as shown in Fig. 2. In this way, one can not know the pressure drop resulting of particle spouting accurately, because the pressure drop between points A&B

also includes pressure drops produced by the additional unit components, such as the nozzle, the insulation panel, the gas buffer device and others. Pressure drops in these specially designed graphite devices are often very large and cannot be ignored, so the pressure drop resulting of particle spouting can not be obtained accurately, but the change of pressure drop values at points A&B at different coating time can be calculated from experimental results.

The pressure changes at points A&B in the coating process of four layers are given in Fig 23. It can be found that the pressure change is different in the coating process of four layers at points A&B. Point B is close to the outlet of the furnace, so the gage pressure at point B is almost not changed and nearly zero. The gage pressure at point A is changed distinctly during the coating process. Period a-b is the rising temperature process. Only argon is used as the fluidization gas. The flow rate is 362 L/min. The pressure at point A is increasing with the temperature. Period b-c is the buffer PyC coating process. In this coating process acetylene is introduced into the spouted bed as reactive gas. Similarly, Period d-e, f-g, h-i is the IPyC, SiC, OPyC coating process respectively. Period c-d, e-f, g-h is the process of adjusting the temperature. Besides, the fluidization gas changes to hydrogen at point f and changes back to argon at point g. The flow rate of argon remains 362 L/min in the whole process except the period f-g, and the temperature at point b and c is the same, so the change of the gage pressure drop ΔP from b to c represents the change of particle properties caused by buffer PyC coated layer. Similarly, the changes of ΔP from d to e, f to g, h to i represent the change of particle properties caused by IPyC, SiC, OPyC coated layer respectively. Period i-j is cooling process after the completion of coating experiments. The pressure at point B is not changed, so the change of pressure drop ΔP from h to i is equal to the change of pressure at point A from h to i. The change of pressure drop ΔP from h to i can be read as 580 Pa from the inset in Fig. 24. In the same way, It can be obtained that the changes of pressure drop ΔP from b to c, d to e, f to g are 367, 412, 705 Pa respectively from experimental results.

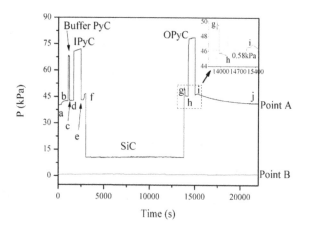

Figure 23. Pressure signals at points A&B during the successive coating process

The minimum spouting velocity can be related with the parameters of the spouted bed geometry and particle properties. Many researchers have given relationships about the minimum spouting velocity in different experiment conditions [19, 20], such as the D. Sathiyamoorthy's equation [21]. It has been indicated that the temperature only influence the constant coefficient and the index number is not changed with temperature. The experiments have been performed in the cold mockup of our spouted bed coating furnace and the following equations are proposed analogy with the literature reports:

$$U_{ms} = k_1 H_0^{1.499} \left(\rho_p - \rho_f \right)^{0.477} d_p^{0.610} \tag{1}$$

The constant k_1 is combination factor of the spouted bed geometrical parameters D_0, D_c, γ fluid density ρ_f and temperature. The pressure drop at stable spouting state ΔP_s can be related to the spouted bed geometry and particle properties, which expressed as follows:

$$\frac{-\Delta P_s}{H_0 \rho_b g} = 1.20 \left(\frac{H_0}{D_0} \right)^{0.08} \left(\text{tg} \frac{\gamma}{2} \right)^{-0.11} \left(\frac{D_0 U_{ms} \rho_f}{\mu} \right)^{-0.06} \tag{2}$$

It can be found that the pressure drop at stable spouting state ΔP_s is a function of the spouted bed geometrical parameters and particle parameters. By substituting Eq.(1) into Eq.(2), considering that the gas density is small, and can be neglected compared to the solid density, then a simplified relationship can be obtained as:

$$-\Delta P_s = k \rho_b H_0^{0.99006} \rho_p^{-0.02862} d_p^{-0.0366} \tag{3}$$

The constant k is a combination factor and is not changed in the same spouted bed under the same temperature, so k is not changed before and after the coating process of each layer. The change of pressure drop can be deduced from Eq.(3). The time points 1 and 2 are represent of the time points before and after the coating process of each layer. Then the simplified form of the pressure drop can be obtained as

$$\frac{-\Delta P_{s1} - \left(-\Delta P_{s2} \right)}{\varphi_1 - \varphi_2} = k, \text{ in which } \varphi = \rho_b H_0^{0.99006} \rho_p^{-0.02862} d_p^{-0.0366} \tag{4}$$

It can be found that, if only particle properties change, for example, in the coating process of particles, the change of pressure drop is linear with the change of combination factor φ about particle density, particle bulk density and static bed height. The change of particle density ρ_p and diameter d_p during the coating process can be calculated, as shown in Fig. 24. Then the bed density ρ_b and static bed height H_0 can be obtained according to the experimental bulk coefficient and the spouted bed geometrical parameters (D_c=150 mm), as shown in Fig. 25. The change of combination factor $\Delta\varphi$ during coating process of buffer PyC, IPyC, SiC, OPyC can be obtained as 28.06, 31.70, 55.83, 45.63 by Eq.(4). The change of pressure drop $\Delta(\Delta P)$ during coating process of buffer PyC, IPyC, SiC, OPyC layer has been obtained as 367, 412, 705, 580 Pa respectively from Fig. 23, k can be calculated as 13.08, 12.99, 12.63, 12.71 in four coating process of buffer PyC, IPyC, SiC, OPyC layer. It can be found that k is almost a constant as shown in Fig. 26. The consistent relationship Eq.(4) is verified by experimental results.

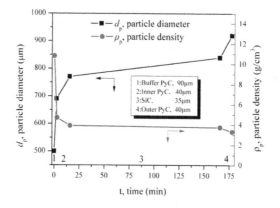

Figure 24. The change in particle diameter and particle density during successive (TRISO) coating on UO₂ micro-spheres

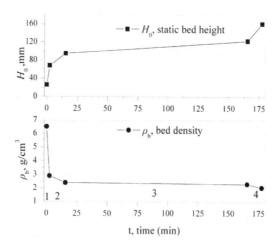

Figure 25. The change in static bed height and bed density during successive (TRISO) coating on UO₂ micro-spheres

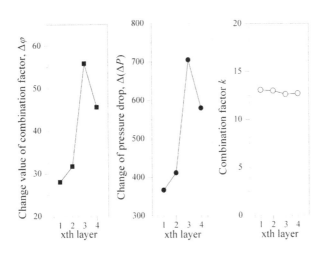

Figure 26. The change of pressure drop, the change of combination factor φ and combination factor k during successive (TRISO) coating on UO_2 micro-spheres

Through the above process, a convenient method for on-line monitoring the fluidized state of the particles in a high-temperature coating process can be proposed. The pressure drop change before and after coating can be used effectively as real-time analysis of particle flow conditions. The combination factor φ can be calculated on the basis of design values of coating layer, the pressure drop change can be obtained from the experimental results, so k can be calculated in four coating process, then a constant k should be obtained. One can compare k values to estimate the fluidization state at different coating process of four coating layers. If an abnormal value is found, which indicates the poor quality of particle fluidization state during the coating process, and then the quality of the coated layer will be affected. Additionally, k can also be estimated firstly by a series of experiments, the pressure drop changes can be calculated by Eq.(4) in the same spouted bed coating furnace according to the design value of the coated layer, and then real-time comparative analysis with the experimental results of pressure drop changes can be performed to monitor the fluidized state of the particles on-line in the coater.

4. Brief summary

The multiscale study of the coating technology of nuclear fuel kernels are discussed above, and the brief summary can be given as follows:

1. In microscale study, it was found that the droplet deposition mechanism can be used to explain the formation of the PyC layer in the fuel particle coating process. The nano-spherical carbon particle is generated at first, and then deposit on the surface of fuel particle. The pure β-SiC can be prepared in the experimental temperature range from 1520-1600ºC. The density change of the SiC coated layer is mainly due to the micropores in the coated layer, not the doped C, Si and other impurities. The oxidization of SiC coating layer was not obvious below 1 200 ºC. The oxidization of SiC coating layer was significant gradually from 1 400 ºC, but the oxidation rate is very slow.

2. In mesoscale study, some key parameters in coating process are proposed, such as single particle cycle time and so on, which can be obtained exactly from DEM-CFD simulation results. The newly proposed gas inlet with swirl flow designed multi-nozzle is better than any other gas inlets for obtaining a more uniform fluidization, which can disperse the gas to increase the gas-particle contact efficiency.

3. In macroscale study, the pressure signals of the coater during the coating process of real TRISO UO$_2$-coated particles are given and analyzed. A relationship about the change of pressure drop and the change of particle properties before and after the coating process of each layer is proposed. This relationship is also validated by the experimental results. A convenient method for real-time monitoring the fluidized state of the particles in a high-temperature coating process is proposed based on the proposed relationship.

5. Conclusions and prospective

In this chapter, the multiscale study concept of the coating technology is stated and validated as an effective and necessary method to develop the coating technology of nuclear fuel kernels from the lab to the factory. It should be indicated that the application of the coating technology for other usages rather than nuclear industry is extensive, such as CNT preparation and suface modification of catalyst particles. The multiscale study of the coating technology can be seen as a universal methodology in the R&D of the coating process, especially in the field of fluidized bed chemical vapor deposition (FB-CVD).

The FB-CVD method is a suitable technique for preparing various kinds of films/layers on the spherical materials by initiating chemical reaction in a gas. This technique can be used for many purposes, such as synthesizing carbon nanotube composite photocatalyst ((CNT)/Fe-Ni/TiO$_2$). Also, some modified method based on FB-CVD, such as plasma-enhanced FB-CVD, has been used to prepare the transparent water-repellent thin films on glass beads in modern surface engineering treatment. So the investigation of FB-CVD method is helpful and impor-tant for modern surface treatments.

This multiscale study should be developed in the future, especially in the the mutual coupling reasearch in micro-, meso- and micro- scales, such as the effect of fluidized state on the homogeneity and compactness of coating materials. Also, this method should play an important role in the study of the scale-up methodology in the field of FB-CVD.

Appendix

γ	Cone included angle
D_o	Inlet diameter, m
D_i	Diameter of the bed bottom, m
D_c	Column diameter, m
H_c	Height of conical part, m
H_o	Static bed height, m
d_p	Particle diameter, m
ρ_p	Particle density, kg/m^3
ρ_f	Fluid density, kg/m^3
ρ_b	Bed density, kg/m^3
ΔP_M	Maximum pressure drop, Pa
ΔP_s	Pressure drop at the stable spouting state, Pa
U_M	Spouting velocity at maximum pressure drop, m/s
U_{ms}	Minimum spouting velocity at the stable spouting state, m/s
g	Gravity, m/s^2
V_p	The volume of spout area, m^3
T_c	The oscillation cycle period of particle clusters, s
T_p	The single particle cycle time, s
T_s	The time period of particle in spout area in a particle cycle time, s

Acknowledgements

The author would like to thank the National S&T Major Project (Grant No. ZX06901), the Independent Research Projects of Tsinghua University (20111080971) and Research Fund for the Doctoral Program of Higher Education of China(20121010010) for the financial support provided. Also, thanks go to Bing Liu, Youlin Shao, Jing Wang, Junguo Zhu, Bing Yang, Bingzhong Zhang, as my best colleagues. All the results of this chapter are completed together with them. The author himself is responsible for all of the errors of this chapter.

Author details

Malin Liu[*]

Address all correspondence to: liumalin@tsinghua.edu.cn

Institute of Nuclear and New Energy Technology, Tsinghua Univerisity, Beijing, China

References

[1] Lee Y-W, Park J-Y, Kim YK, Jeong KC, Kim WK, Kim BG, et al. Development of HTGR-coated particle fuel technology in Korea. Nuclear Engineering and Design. 2008; 238(11): 2842-2853.

[2] Nickel H, Nabielek H, Pott G, Mehner AW. Long time experience with the development of HTR fuel elements in Germany. Nuclear Engineering and Design. 2002; 217(1-2): 141-151.

[3] Sawa K, Ueta S. Research and development on HTGR fuel in the HTTR project. Nuclear Engineering and Design. 2004; 233(1-3): 163-172.

[4] Tang C, Tang Y, Zhu J, Zou Y, Li J, Ni X. Design and manufacture of the fuel element for the 10 MW high temperature gas-cooled reactor. Nuclear Engineering and Design. 2002; 218(1-3): 91-102.

[5] Barnes CM, Marshall DW, Keeley JT, Hunn JD. Results of Tests to Demonstrate a 6-in.-Diameter Coater for Production of TRISO-Coated Particles for Advanced Gas Reactor Experiments. Journal of Engineering for Gas Turbines and Power. 2009; 131(5): 052905.

[6] Reznik B, Gerthsen D, Zhang WG, Huttinger KJ. Microstructure of SiC deposited from methyltrichlorosilane. Journal of the European Ceramic Society. 2003; 23(9): 1499-1508.

[7] Hélary D. , Bourrat X., Dugne O. , Maveyraud G. , Pérez M., P. G. Microstructures of Silicon Carbide and Pyrocarbon Coatings for Fuel Particles for High Temperature Reactors. Proceedings of 2nd International Topical Meeting on HTR Technology. 2004.

[8] Lopez-Honorato E, Tan J, Meadows PJ, Marsh G, Xiao P. TRISO coated fuel particles with enhanced SiC properties. Journal of Nuclear Materials. 2009; 392(2): 219-224.

[9] Liu M, Liu B, Shao Y. The study on pyrolytic carbon powder in the coating process of fuel particle for high-temperature gas-cooled reactor. ICONE18-30137, Volume 1: 545-549. 2010.

[10] Fang C, Liu M-L. The study of the Raman spectra of SiC layers in TRISO particles. Acta Physica Sinica. 2012; 61(9): 097802.

[11] Liu M, Liu B, Shao Y. Study on Properties of SiC Coated Layer for HTGR Fuel Particles (in chinese). Atomic Energy Science and Technology. 2012a; 46(09): 1087-1092.

[12] Lopez-Honorato E, Meadows PJ, Xiao P, Marsh G, Abram TJ. Structure and mechanical properties of pyrolytic carbon produced by fluidized bed chemical vapor deposition. Nuclear Engineering and Design. 2008; 238(11): 3121-3128.

[13] Rohmfeld S, Hundhausen M, Ley L. Raman scattering in polycrystalline 3C-SiC: Influence of stacking faults. Physical Review B. 1998; 58(15): 9858-62.

[14] Liu M, Liu B, Shao Y. Optimization of the UO2 kernel coating process by 2D simulation of spouted bed dynamics in the coater. Nuclear Engineering and Design. 2012b; 251: 124-130.

[15] Duarte CR, Olazar M, Murata VV, Barrozo MAS. Numerical simulation and experimental study of fluid-particle flows in a spouted bed. Powder Technology. 2009; 188(3): 195-205.

[16] Lopes NEC, Moris VAS, Taranto OP. Analysis of spouted bed pressure fluctuations during particle coating. Chemical Engineering and Processing. 2009; 48(6): 1129-1134.

[17] Xu JA, Bao XJ, Wei WS, Shi G, Shen SK, Bi HT, et al. Statistical and frequency analysis of pressure fluctuations in spouted beds. Powder Technology. 2004; 140(1-2): 141-154.

[18] Liu M, Shao Y, Liu B. Pressure analysis in the fabrication process of TRISO UO2-coated fuel particle. Nuclear Engineering and Design. 2012c; 250: 277-283.

[19] Zhong WQ, Chen XP, Zhang MY. Hydrodynamic characteristics of spout-fluid bed: Pressure drop and minimum spouting/spout-fluidizing velocity. Chemical Engineering Journal. 2006; 118(1-2): 37-46.

[20] Zhou J, Bruns DD, Finney CEA, Daw CS, Pannala S, Mccollum DL. Hydrodynamic Correlations with Experimental Results from Cold Mockup Spouted Beds for Nuclear Fuel Particle Coating. AIChE Annual Meeting. 2005.

[21] Sathiyamoorthy D, Rao VG, Rao PT, Mollick PK. Development of pyrolytic carbon coated zirconia pebbles in a high temperature spouted bed. Indian Journal of Engineering and Materials Sciences. 2007; 17(5): 349-52.

Chapter 8

Electrodeposition of Alloys Coatings from Electrolytic Baths Prepared by Recovery of Exhausted Batteries for Corrosion Protection

Paulo S. da Silva, Jose M. Maciel, Karen Wohnrath, Almir Spinelli and Jarem R. Garcia

Additional information is available at the end of the chapter

1. Introduction

Electrical and electronic equipment have developed rapidly and their average life spans have been reduced due to the changes in functions and designs [1-3]. Recently, the recovery of precious metals from these electronic scraps has become attractive. Precious metals and copper in PC board scraps and waste mobile phones account for more than 95% of the total intrinsic value [4] and recently several authors are carrying out study on the applicability of economically feasible hydrometallurgical processing routes to recover precious metals [3-5].

Due to this massive industrialization of electronic equipment like toys, cameras, laptops, cell phones, etc [6]. In recent years there has been a considerable increase in the consumption of household batteries The American industry invoice approximately 2.5 billion dollars annually selling about 3 billion batteries. In Europe in the year of 2003 were produced 160,000 tonnes of portable batteries. In this year in Brazil the annual production of these devices reaches about 1 billion units [7, 8]. In this way spent batteries represent an increasing environmental problem due to the high content of heavy metals. Unlike large batteries used for vehicles, small, portable batteries are very diverse in terms of chemical composition and represent 80–90% of all portable batteries collected [9, 10]. The difference between various types of used battery is represented by the used materials such as electrolytes and electrodes [9]. These batteries can be sorted by size, shape and chemical composition so that we can determine which metals can be recovered from each category.

For these reasons in several countries, collecting batteries is becoming mandatory, and so is recycling those containing toxic materials. Recycling may also be applied to recovering valuable materials to be reutilized [11].

There are basically two types of household batteries: primary batteries that after becoming worn are discarded and the secondary batteries that can be recharged [12, 13]. Within the wide range of commercially available batteries zinc-carbon batteries (also known as Leclenché or dry cells) and alkaline batteries are the most consumed because of its low cost. In Europe, from the total of batteries sold in 2003, 30.5% and 60.3%, were Zn-C batteries and alkaline batteries, respectively. In China are produced annually more than 15 billion of these devices and in Brazil estimating a consumption of six batteries per inhabitant per year [7, 8, 14].

The disposal of these batteries is a serious problem, because in their composition there are metals considered dangerous to the environment [13]. The cost for the safe disposal of these materials is quite high due to the large amount of dangerous waste generated and due to the fact that the storage capacity in landfills or dumps is running out. A policy adopted in 2006 by the European Union (EU) banned incineration and disposal of batteries in landfills. This regulation applies to all types of batteries regardless of shape, volume, weight, composition or use. Through this new policy it is expected to mobilize the EU countries member for the collection, recovery and recycling of metals present in these power devices [15]. In Brazil, according to the resolution 401/2008 of the *Brazilian National Council of the Environment* (CONAMA in Portuguese) [16], after consumption, household batteries must be collected and sent to the manufacturers, to be recycled, treated or disposed of an environmentally safe way, but until 1999 they could be disposed of in household waste since meet the limits of heavy metals in its composition. Although required, this resolution proved to be insufficient to solve the problem of environmental contamination by means of this waste since there is a large annual consumption of these batteries. A factor to be noted is that in spite of Zn and Mn match most of the composition of cells Zn-MnO$_2$, the limits of contamination of these metals are not established by law. Another aggravating factor is the use of irregular cells entering the Brazilian market. Frequently these products do not meet manufacturing standards. The heavy metal content of these cells is seven times greater than that limited established by the CON-AMA. Thus, the contamination starts by improper disposal of these devices in landfills or dumps, which is the destination of the majority of household solid waste in Brazil [13,16, 17].

Industrial recycling of batteries is generally focused on two processes: the pyrometallurgical and/or the hydrometallurgical. The pyrometallurgical method is based on the difference of volatilization of different metals at high temperatures followed by condensation. The hydro-metallurgical method is based on the dissolution of metals in acidic or alkaline solutions. The advantage of the first method is the absence of the necessity of dismantlement of the devices. However, it is an expensive process, since it requires high temperatures and is not efficient selectively, for example, to obtain pure zinc from Zn-MnO$_2$ batteries, Ni-Cd batteries cannot be treated simultaneously because the Zn and Cd are not selectively volatized in the oven, so sorting steps are required in advance of the materials recycling. Another drawback is related to the production of dust and gas emission into the atmosphere during the recycling process. The hydrometallurgical route is usually more economical and efficient than the pyrometallurgical process. In addition there is a diminishment of the emission of particles into the atmosphere. However, it is a more laborious process requiring pretreatment steps such as sorting, disassembling and leaching of material to improve the dissolution of metals in aqueous phase. In addition, the recovery of metals requires different aqueous media (acid or alkaline) and various processes of precipitation [14, 18, 19].

The recovery of these materials is very important because in addition to reducing the enormous amount of waste generated by consume of these power devices, recycling could lower the cost of production of new batteries through the reuse of raw materials by recycling and, consequently it reduces the risks to the environment [19, 20].

2. Features of Zn-MnO$_2$ batteries

Zn-C (also known as Leclanché or Zn-MnO$_2$ Batteries) batteries and alkaline batteries are basically composed by potassium, manganese and zinc as metal species. The stack of Zn-C was invented in 1860 by George Leclanché and the devices currently used are very similar to the original version.

A schematic view of this type of batteries is showed in Figure 1. In these batteries the anode consists of a zinc metal cylinder used, usually in the form of plate to procedure the outside structure of the cell. The cathode consists of a graphite rod surrounded by a powder mixture of graphite and manganese dioxide. The electrolyte is a mixture of ammonium chloride and zinc chloride. During the Zn-C and alkaline batteries discharge, basically the following reactions are observed:

Zinc oxidation at anode:

$$Zn + 2NH_4Cl + 2OH^- \rightarrow Zn(NH_3)_2 Cl_2 + 2H_2O + 2\ e^- \tag{1}$$

Manganese reduction at cathode:

$$2MnO_2 + 2H_2O + 2e^- \rightarrow 2OH^- + 2MnOOH \tag{2}$$

Resulting in the overall reaction:

$$Zn + 2MnO_2 + 2NH_4Cl \rightarrow Zn(NH_3)_2 Cl_2 + 2MnOOH \tag{3}$$

In this kind of batteries, during storage and in rest periods while operating some parallel reactions can occur, causing leaks and loss of efficiency. In this way, some metals such as Cd,

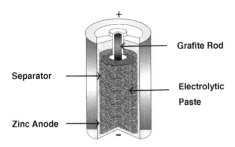

Figure 1. Schematic View of the Zn-MnO₂ Battery (Leclanché Device)

Cr, Hg and Pb are added to these devices to improve their performance and to avoid these parallel reactions.

The alkaline battery is a modified version of the stack of Zn-C. It features the same electrodes (anode and cathode), however, the electrolyte is a concentrated potassium hydroxide folder containing zinc oxide. Another difference is that its outer part is made on steel plate for assuring better seal. The reactions that occur in the cathode during discharge are the same that occurs in the Zn-C batteries, but the anodic reactions are different:

Zinc oxidation at alkaline batteries anode:

$$Zn + 2OH^- \rightarrow Zn(OH)_2 + 2e^- \tag{4}$$

Resulting in the overall reaction:

$$Zn + 2MnO_2 + 2H_2O \rightarrow Zn(OH)_2 + 2MnOOH \tag{5}$$

The advantage of the alkaline batteries is that they do not have parallel reactions and can be stored for up to four years keeping more than 80% of their original capacity, additionally its lifetime is up to ten times higher, however they are on average five times more expensive. Alkaline batteries are placed on the market as "mercury-free", however the literature reports that in several works on recycling of batteries were found heavy metals in these devices, including mercury [19-23]. The composition of some alkaline batteries and Zn-C are given in Table 1.

These portable batteries (alkaline and Zn–C) contain Mn and Zn in high concentrations. Considering that the production of these kind of power device are increasing it has become important the usage of recycled metals production instead of primary metals. Besides the positive impact on the environment, in the recuperation process of materials lot of energy is

Metal	A1[a] % in weight*	A2[b] % in weight*	B1[c] % in weight*	B2[d] % in weight*
Zn	21	20,56	5	5,05
Mn	45	26,60	23-30	29,04
Fe	0,36	0,15	0,2-10	0,18
Hg	1 (ppm)	0,0012	-	-
Cd	0,06(ppm)	0,0007	-	0,0002
Pb	0,03	0,005	-	-
Ni	-	0,008	0,007	0,006
K	4,7	7,3	-	-

* (% in weight of the electrolytic paste)

[a]Alkaline [23]; [b]Alkaline [22]; [c]Zn-C [23] and[d]Zn-C [21].

Table 1. Composition of $Zn-MnO_2$ and Alkaline Cells.

saved and the pollution is also reduced as the chemical treatment of primary metals is not needed. Manganese and zinc are important metals in many fields. Zinc is the most important nonferrous metal after copper and aluminum [23] and of the total zinc consumption, 55% is used to cover other metals to prevent oxidation, 21% in zinc-based alloys, 16% in brass and bronze. The increase of zinc demand in 2010 was due to a revival of the consumption in Europe (24%) and also to the consolidated economic growth of the emerging economies like Brazil, India and most notably China where the consumption increased 11% respect to 2009. Most of consumption of manganese is related to steel production, directly in pig iron manufacture and in the ferroalloy industry. Manganese resources are large but irregularly widespread in the world and South Africa and Ukraine account for about 75% and 10 % of the word's identified manganese resources respectively [24].

Due to the growing interest in global environmental issues, recycling of Zn–Mn batteries carried more attentions and was reviewed in detail recently [15]. As the most widely used hydrometallurgical process, acid leaching was frequently used to release both Zn and Mn from the spent Zn–Mn batteries in the presence of strong acid solution such as H_2SO_4, HCl, HNO_3 and so on. In most cases, acid leaching produce nearly 100% of Zn extraction from the spent batteries, but Mn dissolution was rather poor due to insoluble MnO_2; moreover, the heavy consumption of various strong acids endowed the leaching process with high cost, strict requirements of equipment and potentially safe risk. The reductive acidic leaching could greatly improve extraction yield of Mn by adding inorganic reductants such as H_2O_2 and SO_2 or organic ones such as glucose, sucrose, lactose, oxalic acid, citric acid, tartaric acid, formic acid and triethanolamine, but higher safety risk and greater operation cost occurred [15]. So, developing the environmentally-friendly and cost-effective recycling methods for the spent Zn–Mn batteries are encouraged. At present, biohydrometallurgical processes (bioleachingtech) have been gradually replacing hydrometallurgical ones due to their higher efficiency, lower cost and few industrial requirements [25]. Bioleaching was characterized by efficient release of metals from solid phase into aqueous solution under the mild conditions of room temperature and pressure by contact and/or non-contact mechanisms in the presence of acidophilic sulfur-oxidizing and/or iron-oxidizing bacteria [26, 27].

Another alternative method developed for Zn-Mn batteries recycling is the electrodeposition of Zn and Zn–Mn alloy coatings over different kinds of steel to corrosion protection [28, 29]. Electrodeposited coatings of zinc are extensively employed in the protection of steel against corrosion. However, this protective effect is not very effective under aggressive atmospheric conditions [30]. In recent years, several materials have been investigated to improve the durability of these coatings. Electrodeposited alloys of Zn, such as Zn–Ni, Zn–Co and Zn–Fe, present higher corrosion resistance than pure zinc coatings. Also, it has been reported in the literature that Zn–Mn alloys show even better corrosion resistance properties [31–34]. The high corrosion resistance of these alloys is likely due to the dual protective effect of manganese: on the one hand Mn dissolves first because it is thermodynamically less noble than Zn, thereby protecting Zn; and on the other hand Mn ensures the formation of compounds with a low solubility product over the galvanic coating. Depending on the aggressivity of the environment to which the Zn–Mn alloy is exposed, various compounds may be found in the passive layer, including oxides such as MnO, MnO_2, Mn_5O_8 and γ–Mn_2O_3, or basic salts like $Zn_4(OH)_6SO_4.xH_2O$ and $Zn_5(OH)_8Cl.2H_2O$ [31, 35, 36]. The protective effect of Zn–Mn is dependent on the Mn content of the alloy. Although it has been reported that among the Zn alloys those of Zn–Mn show the highest corrosion resistance, their deposition process presents some drawbacks related to the bath instability and current efficiency. Among the various electrolytic baths and additives proposed to obtain Zn–Mn alloys, the use of a chloride-based acid bath with polyethylene glycol (PEG) as the additive seems very promising [31–33].

The mainly practical application in produce a protective Zn-Mn layer over steel is related to the substitution of the primary painting process on metallic parts produced in foundries. Furthermore it is important to note that beside the great interest in recycling Zn-C batteries the use solution produced by the acidic leaching of these exhausted batteries to obtaining protective Zn-Mn films were described only in two papers[28,29].

Considering the concepts described above this chapter brings some highlighting on the development of a methodology to recover zinc and manganese present in exhausted zinc–carbon batteries through chloride acidic leaching of the solid material. The leaching solution is then used as an electrolytic bath for the electrodeposition of the galvanic coating on AISI 1018 steel. Polyethylene glycol is used as the additive in the bath to obtain both Zn and Zn–Mn alloys.

3. Electrodeposition of Zn and Zn–Mn alloy coatings

To carrying out the electrodeposition of Zn and Zn–Mn Alloy coatings from electrolytic baths obtained from leaching of spent batteries is important to know the metallic composition and concentration of the solution. Table 2 shows the typical composition obtained by an acid leaching with diluted HCl, of the carbon paste obtained from Zn-C spent batteries manufactured in Brazil [28]. This quantitative analysis of the metal content was performed by atomic absorption spectrometry and indicated that in addition to Zn^{2+} and Mn^{2+} traces of other species such as Fe^{2+}, Cu^{2+} and Pb^{2+} were present in solution.

Metal	Concentration (mg L⁻¹)	Metal	Concentration (mg L⁻¹)
Zn	6,981.33	Pb	55
Mn	3,030.30	Cr	0
Cu	30.4	Ag	0
Fe	5.5	Ni	0

Table 2. Composition of the Electrolytic Bath Obtained from Recycling of Zn-C Batteries.

It could be seen from the Table above that the batteries used containing Pb as heavy metal in their composition, however this quantity found corresponds to 0.18% by weight of the battery electrolytic paste that was in agreement with the limits established by CONOMA (0.20%).

To prepare the electrolytic bath from this solution the pH is adjusted to 5.0. During this step occurs the hydrolysis of some metallic ions forming a gelatinous brown material, possibly due to the formation of iron hydroxide that was removed by filtration. From this filtrated solution it was prepared four electrolytic baths used in obtaining the Zn-Mn alloy coatings. These baths were prepared to the addition of boric acid and different quantities of additives (ammonium isocyanate - NH_4SCN and polyethylene glycol – $PEG_{10.000}$) as showed in Table 3.

Bath Name	Zn	Mn	PEG₁₀.₀₀₀	NH₄SCN	H₃BO₃
S₀			-	-	
S₁	0.10 mol L⁻¹	0.06 mol L⁻¹	1 g L⁻¹	-	0.32 mol L⁻¹
S₂			-	6.5 mmol L⁻¹	
S₃			1 g L⁻¹	6.5 mmol L⁻¹	

Table 3. Composition of the Obtained Electrolytic Bath through Recycling of Cells Used to Obtain Mn-Zn Alloys.

The behavior of AISI 1018 carbon steel electrodes in the presence of the electrolytic baths prepared from recycled batteries could be investigated by measurements of cyclic voltammetry. The Figure 2 shows the voltammetric curves obtained on 1018 carbon steel immersed in the proposed electrolytic baths (see Table 3). It could be seen from Figure 2a that with no additive on the bath the voltammogram showed two regions of reduction. The first region present a peak current with maximum current in -1.4 V and is related to the electrodeposition of zinc on the electrode. The second region present a continuous increase on reduction current starting on E = -1.5 V, this region can be related to the formation of Mn-Zn alloy, but tis increase in cathodic current has an important contribution of the process of hydrogen evolution. This former observation is the reason of the needing of usage of additives that could be a hindrance for this reaction on the surface. When the potential is swept in the positive direction no peak of current is observed, just a constant increase on anodic current starting in almost E = -1.2 V, this current increase is related to the dissolution of metal layer deposited on the steel during the cathodic scan.

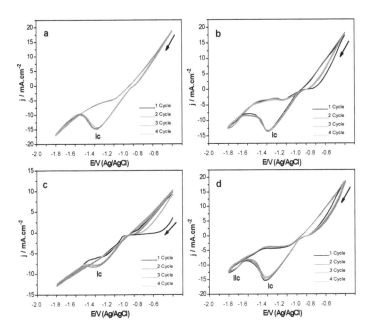

Figure 2. Voltammetric Curves Obtained on 1018 Carbon Steel Immersed in the Proposed Electrolytic Baths (see Table 3) (a) S0, (b) S1, (c) S2 and (d) S3 with scan rate = 20 mV s^{-1}.

No significant changes are observed with the addition of the addictive PEG$_{10,000}$ and with the addition of a mixture of NH$_4$SCN and PEG$_{10,000}$ (see Figures 2b and 2d), except by a slight diminishment on the current density, this loss in current density could be assigned to the presence of the additives. As discussed by Diaz-Arista et al. [31] the PEG$_{10,000}$ can adsorb on the steel surface blocking of the active sites for hydrogen evolution, but the isocyanate, according to these authors has the function of complexing with the ions Zn^{2+} decreasing the competition with the reduction of ions Mn^{2+}. However, in the presence of NH$_4$SCN the voltammetric curves are considerably different from the others conditions, as could be seen in Figure 2c. This figure shows that the process of Zn^{2+} electrodeposition, that occurred with a peak current at ca. -1.4 V in the other conditions, is now linked to the Mn^{2+} electrodeposition process as could be observed by the continuous increase in the cathodic current with the augment of the potential forward negatives values. This behavior was assigned to the fact that the NH_4SCN has complex interactions with Zn^{2+} and Mn^{2+} decreasing the difference between the reduction potential of these two metallic species on the steel surface. The characterization of the coatings obtained with NH_4SCN as addictive showed films with bad quality as weak adherence to the surface, inhomogeneity, and with no increase in the Mn proportion related to the presence of Zn. For this reason this additive was not used in the others experiments described in this text.

After determining the potential range for reduction of the metallic ions Zn^{2+} and Mn^{2+} presented in the prepared electrolytic baths, the coating were obtained in the potentiostatic mode in two situations: first by application of -1.2 V and the second by application of -1.6 V during 15 min. The conditions used during the potentiostatic electrodeposition using the S0 and S1 solutions were chosen in agreement with previous studies carried out by other authors on the electro-deposition of Zn–Mn alloys [31]. The resulting current–time curves are shown in Figure 3. At -1.2 V, in the absence of additive, the current density stabilized at around -4.5 mA cm^{-2}. In the presence of PEG, the current density is slightly less negative (-3.6 mA cm^{-2}). When electrode-position was carried out at -1.6 V the current densities with and without the additive were -17.6 mA cm^{-2} and -10.8 mA cm^{-2}, respectively. It is important to note that the presence of additive during the electrodeposition at more negative potentials makes the current density more stable. The instability observed in the electrodepositions carried out with S0 solution, particularly at -1.6 V, may be attributed to hydrogen evolution. In a study on the effect of additives on the hydrogen evolution reaction during Zn electrodeposition, Song et al. [37] suggested that PEG acts as an inhibitor of hydrogen absorption in the electrodeposited Zn.

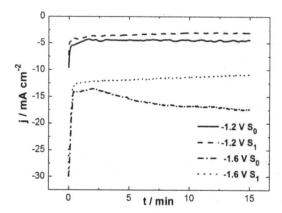

Figure 3. Potentiostatic curves obtained during electrodeposition onto an AISI 1018 steel electrode from S0 and S1 baths.

The characterization of the coating obtained potentiostatically was performed by measure-ments of Scan Electronic Microscopy (SEM), Energy Dispersion Spectroscopy (EDS) and X Ray Diffraction (XRD). The Figure 4 shows the morphology of the deposit obtained potentiostati-cally at -1.2 V from the base solution (S0). The SEM image shows that the deposit is comprised of hexagonal plates with pyramidal clusters grouped into nodules of several sizes, as is normal for pure zinc electrodeposits [37]. The EDS analysis (inset on Figure 4) showed the presence of zinc as the predominant element in the coating.

In the presence of PEG, the deposit obtained is comprised of hexagonal crystals oriented perpendicularly to the substrate surface (Figure 5) and the zinc also was the predominant element (inset on Figure 5). This type of morphology was also observed by Ballesteros *et al.* [30] who studied the influence of PEG as an additive on the mechanism of Zn deposition and nucleation.

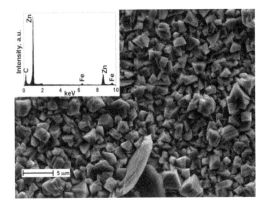

Figure 4. SEM image of the deposit formed on AISI 1018 steel electrode at -1.2V vs. (Ag/AgCl), polarization time = 15 min, using solution S0 as electrolytic bath. The results for the EDS analysis of the film are shown in the inset.

Figure 5. SEM image of the deposit formed on AISI 1018 steel electrode at -1.2V vs. (Ag/AgCl), polarization time = 15 min, using solution S1 as electrolytic bath. The results for the EDS analysis of the film are shown in the inset.

Although differences were observed in the morphology of the deposited coatings obtained with and without the use of the additive, the XRD analysis exhibited in Figure 6 showed the characteristics diffraction peaks for the coating obtained on AISI 1018 steel electrode at −1.2 V vs. (Ag/AgCl), t = 15 min, using the solution S0 (without additive). A similar composition was obtained with the bath S1 in the same electrodeposition condition. As can be seen, the formation of a Zn-Mn alloy could not be obtained from this potentiostatic experiments carried out at -1.2 V.

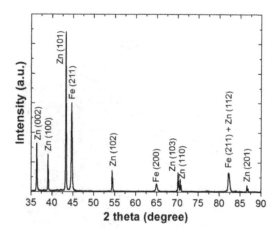

Figure 6. X-ray diffraction (XRD) pattern of the deposit obtained on AISI 1018 steel electrode at −1.2 V vs. (Ag/AgCl), t = 15 min.using the solution S0 (without additive).

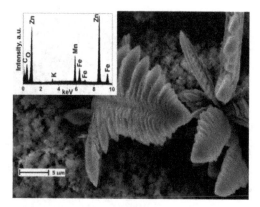

Figure 7. SEM image of the deposit formed on AISI 1018 steel electrode at −1.6 V vs. (Ag/AgCl), t = 15 min, from solution S0. The results for the EDS analysis are shown in the inset.

Figure 7 shows the morphology of the deposit obtained potentiostatically at -1.6 V without the use of the additive. In the SEM image an amorphous and porous deposit covering some parts of the substrate can be observed. The EDS analysis (inset on Figure 7) indicated that the manganese content of this deposit is around 8% wt.

The XRD analysis of the coating obtained at −1.6 V vs. (Ag/AgCl) from solution S0 is showed in Figure 8. This XDR measurement also indicated that these experimental conditions did not favor the formation of a Zn-Mn alloy. This result may be related to the formation of Mn(OH)$_2$(s) species on the substrate surface due to the hydrogen formation under these experimental conditions, resulting in an increase in the pH in the vicinity of the working electrode [38]. In addition, the formation of Mn(OH)$_2$ in high alkaline conditions agrees well with the Eh x pH (Pourbaix) diagrams [39].

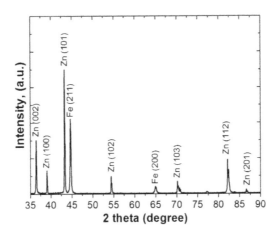

Figure 8. X-ray diffraction (XRD) pattern of the deposit obtained on AISI 1018 steel electrode at −1.6 V vs. (Ag/AgCl), t = 15 min.using the solution S0 (without additive).

In comparison with the morphology observed for the deposit obtained without PEG, the deposit formed in the presence of this additive is very different. The SEM image and EDS analysis for this coating are showed in Figure 9.

This figure shows that the deposit formed is compact and homogeneous with a cauliflower-like morphology; however, once again, the presence of manganese in the deposit could not be detected. The change in the morphology of the deposit may be associated with the partial adsorption of the additive on the electrode surface during the electrodeposition of Zn^{2+} [40]. In addition, the XRD analysis exhibited in Figure 10 revealed that the deposit is comprised principally of Zn crystals in the plane (101), corroborating the EDS results.

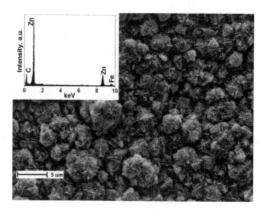

Figure 9. SEM image of the deposit formed on AISI 1018 steel electrode at −1.6 V vs. (Ag/AgCl), t = 15 min, from solution S1. The results for the EDS analysis are shown in the inset.

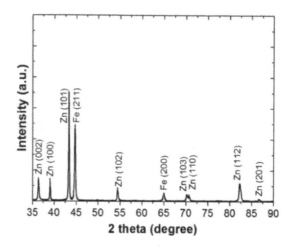

Figure 10. X-ray diffraction (XRD) pattern of the deposit obtained on AISI 1018 steel electrode at −1.6 V vs. (Ag/AgCl), t = 15 min, using the solution S1 (with PEG$_{10,000}$ as additive).

As it was not possible to obtain films containing Zn-Mn alloy through potentiostatic electro-deposition, coatings were obtained by galvanostatic deposition. As the current density stabilized at around -10 mA cm^{-2} with the presence of PEG in the electrolytic bath during potentiostatic electrodeposition at -1.6 V, this current density was chosen for the attempted galvanostatic electrodeposition of the Zn-Mn alloy. Figure 11 shows the chronopotentiometric curves obtained.

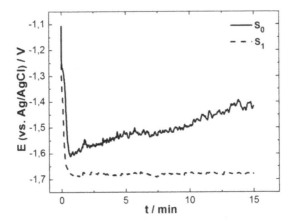

Figure 11. Chronopotentiometric curves obtained during electrodeposition of the deposits on AISI 1018 steel electrode from base solution S0 and solution S1.

In the absence of PEG the potential changed during the electrodeposition, resulting in a rough and irregular deposit, as evidenced in the SEM analysis. On the other hand, when the additive was added to the base solution, the deposition potential stabilized at around -1.68 V at the beginning of the electrodeposition. Additives such as PEG can shift the potential of Zn deposition to more negative values, enabling Zn alloys to be obtained with metals for which the deposition potentials are very negative [30]. In addition, the use of PEG as an additive allowed a compact and homogeneous deposit to be obtained.

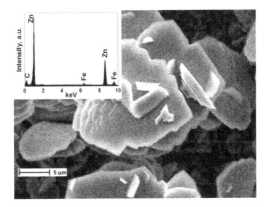

Figure 12. SEM image of the deposit formed on AISI 1018 steel electrode at −10 mA cm^{-2}, t = 15 min, from solution S0. The results for the EDS analysis are shown in the inset.

Figure 12 shows the SEM image of the deposit obtained galvanostatically at -10 mA cm^{-2} in the absence of PEG. The deposit formed is porous and with grains of diverse dimensions irregularly distributed on the substrate surface. The EDS analysis (inset in Figure 12) revealed that there is no manganese present in the coating.

In addition, the XRD patterns of the film showed above were very similar as those obtained during the potentiostatic deposition at -1.2 V from solution S0n (see Figure 6) indicating only the presence of the Zn crystals in different planes.

The results obtained in the presence of PEG indicated the formation of the Zn-Mn alloy during galvanostatic electrodeposition. Figure 13 shows the SEM micrograph of the deposit obtained under these conditions. The results indicate that the PEG decreased the mean size of grains inducing the formation of a smooth deposit. Although the EDS analysis did not clearly indicate the presence of Mn in the deposit.

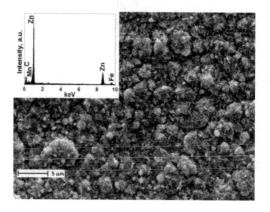

Figure 13. SEM image of the deposit formed on AISI 1018 steel electrode at −10 mA cm^{-2}, t = 15 min, from solution S1. The results for the EDS analysis are shown in the inset.

The XRD results, exhibited in Figure 14 showed a diffractogram characteristic of a mixture of Zn and ε-phase Zn-Mn with different crystallographic. This finding may be related to the low manganese content, around 2% wt in the deposit. Ballesteros et al.[30] have reported that the presence of additives such as PEG can shift the potential of Zn deposition to very negative values. Such behavior is associated with the partial adsorption of PEG onto the substrate surface. The authors related that in the presence this additive the electrodeposition of zinc can occurs in two different ways. First, the zinc is electrodeposited onto the active sites on the electrode surface that are not blocked by adsorbed PEG molecules. In second, the zinc is electrodeposited onto the active sites that are liberated when PEG molecules are desorbs from electrode surface. This occurs in potential very more negative than the first. The effect of displacement of the zinc reduction potential to more negative values is known as cathodic polarization. Accordingly, potentials as negative as -1.6V vs. SCE could be used to obtain deposits of zinc alloys with metals such as Mn. Similar XRD results were observed by Sylla et *al.* [32]. They reported a Mn content of 1% wt in a Zn-Mn alloy deposit obtained from a chloride-based acidic bath containing PEG as an additive. The authors postulated that the presence of PEG allowed the formation of a compact and homogeneous deposit with cauliflower-like morphology. However, the presence of PEG in the solution hindered manganese deposition and inhibited the formation of the ζ-phase Zn-Mn. It is important to note that the peaks observed in our study for Zn and the phases of Zn-Mn alloy are very close and some overlap may have occurred.

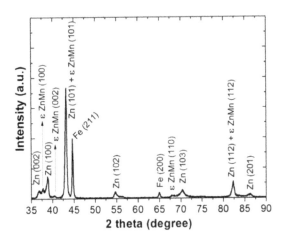

Figure 14. X-ray diffraction (XRD) pattern of a deposit obtained on AISI 1018 steel electrode at -10 mA cm^{-2}, t = 15 min from solution S1 (with PEG$_{10,000}$ as additive)

Table 4 shows a summary of all parameters used in the electrodeposition and some characteristics of the deposits obtained.

Without additive Deposit		
-1.2 V	Homogeneous, comprised by hexagonal plates	only Zn
-1.6 V	Amorphous and porous, Zn and Mn(OH)$_2$	(~ 8% w/t Mn)
-10 mA cm^{-2}	Amorphous and porous	only Zn
With additive Deposit		
-1.2 V	Homogeneous , comprised by hexagonal crystals	only Zn
-1.6 V	Homogeneous, cauliflower morphology	only Zn
-10 mA cm^{-2}	Homogeneous, smooth	Zn, Zn-Mn alloys, (~2% w/t Mn)

Table 4. Electrodeposition parameters and characteristics of the deposits obtained

The evaluation of the corrosion resistance of the coating obtained could be performed by measurements of polarization curves. Figure 15 shows the polarization curves of the produced coatings in a solution containing NaCl 3% (w/v).

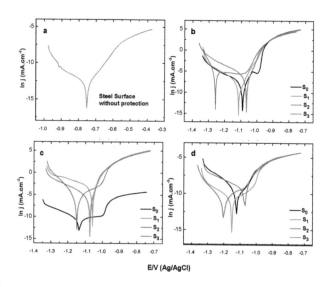

Figure 15. Potentialdynamic Polarization Curves (a) steel surface without protection (b) coating obtained potentiostaticly at -1.2 V vs (Ag/AgCl$_{sat.}$); (c) coating obtained potentiostaticly at -1,6 V vs (Ag/AgCl$_{sat.}$) and (d) coating obtained galvanostaticly at -10mA.cm^{-2}.

The results showed above include the measurements of the coatings obtained using NH$_4$SCN as additive to illustrate the poor protection of the coating produced with this additive due to the lack of homogeneity as described above (see Table 3). The curves showed in Figure 15 indicate that in general, all coatings shift the corrosion potential value (E$_{corr}$) to negative regions when compared with the E$_{corr}$ of the substrate (-0.77 V), this behavior is characteristic for the formation of cathode coatings. It could be also noted from Figure 15 that the corrosion current density (i$_{corr}$) is relatively higher for the coating that presented more negative values of E$_{corr}$. This fact could be seen as divergences but it needs to take in account that the coating that presented more negative values for E$_{corr}$ presented also higher roughness, fact that increase their superficial area causing this increase in current.

Almost at the same time that the work described in this text was done, others experiments using a very similar approach were done by Brito et al. [29]. These authors used H$_2$SO$_4$ solution to leaching spent Zn-C and used methylamine as addictive. They found that the quality of deposits produced from lixiviation depends strongly on the magnitude of the electrodeposition current; homogeneous and uniform deposit layers with good anticorrosive properties were obtained, preferentially, at low current densities. The Mn/Zn mass ratios in the produced deposit layers are influenced by electrodeposition currents and the electrodeposition duration.

Lower electrodeposition currents and shorter electrodeposition duration improve the deposition of Mn in relation to Zn. The presence of the methylamine, also benefit the deposition of Mn and that the addition of methylamine to the electrodeposition baths contributes to the establishment of deposit coatings, with better anticorrosive properties.

4. Conclusions

The results reported by our group[28] and by the group of the Portalegre Polytechnical Institute - Portugal[29] demonstrate that it is possible to obtain galvanic coatings in a bath prepared from zinc and manganese recovered from exhausted zinc-carbon batteries. Additionally these showed that the production of a protective Zn-Mn layer over steel related to the substitution of the primary painting process on metallic parts produced in foundries is possible to be developed as practical application.

Specifically thought in the results obtained in the experiments described it could be observed that the presence of polyethylene glycol or methylamine as additives in the electrolytic bath during electrochemical deposition favors the obtainment of a compact and homogeneous deposit containing a mixture of Zn and a Zn-Mn alloy, with a manganese content in the range of 2% to 7% in weight and the electrodeposited coatings with higher Mn content improve the anticorrosiveness of mild steel in saline environments. The proposed method may represent an alternative use for zinc and manganese recovered from exhausted alkaline and zinc-carbon batteries and thus minimize the adverse environment impacts caused by these residues. Moreover lixiviation solutions resulting from the hydrometallurgical treatment of spent domestic batteries, mainly, $Zn-MnO_2$ batteries can be valued directly as electrodeposition baths for zinc and zinc alloys.

The fact that the methodology developed for recycling batteries can now produce the separation of the elements of which these batteries are made can open new avenues of application for these recovered materials. As an example it could be produced nanostructures of Zinc Oxide by electrodeposition over different kind of surfaces that could be used as quantum dots in photovoltaic devices. Also it could be produce magnetic films from the hydrometallurgical treatment of spent $Zn-MnO_2$ batteries by the electrochemical deposition of ferrites with different contents of Zn and Mn in the structure. However these subjects are still possibilities, because no work involving these potentials could be found in the literature.

Acknowledgements

The authors are grateful for the financial assistance provided by CAPES, CNPq and Fundação Araucária (Brazil).

Author details

Paulo S. da Silva[1], Jose M. Maciel[2], Karen Wohnrath[2], Almir Spinelli[1] and Jarem R. Garcia[2*]

1 Chemistry Department, Federal University of Santa Catarina, Florianópolis, SC, Brazil

2 Chemistry Department, State University of Ponta Grossa, Ponta Grossa, PR, Brazil

References

[1] Naseri Joda NRashchi F. Recovery of Ultra-Fine Grained Silver and Copper from PC Board Scraps. Separation and Purification Technology (2012). , 92-36.

[2] Park, Y. J, & Fray, D. J. Recovery of high purity precious metals from printed circuit boards, Journal of Hazardous Materials (2009). , 164-1152.

[3] Hageluken, C. Improving metal returns and eco-efficiency in electronics recycling, in: Proceedings of the 2006 IEEE International Symposium on Electronics and the Environment, San Francisco, USA, (2006). , 218-223.

[4] Cui, J, & Zhang, L. Metallurgical recovery of metals from electronic waste: a review, Journal of Hazardous Materials (2008). , 58-228.

[5] Quinet, P, Proost, J, & Van Lierde, A. Recovery of precious metals from electronic scrap by hydrometallurgical processing routes. Minerals and Metallurgical Processing (2005). , 22(1), 17-22.

[6] Ruffino, B, Zanetti, M. C, & Marini, P. A mechanical pre-treatment process for the valorization of useful fractions from spent batteries. Resources, Conservation and Recycling (2011). , 55-309.

[7] Bernardes, A. M, Espinosa, D. C. R, & Tenório, J. A. S. Collection and recycling of portable batteries: a worldwide overview compared to the Brazilian situation. Journal of Powder Sources (2003). , 124-586.

[8] Salgado, A. L, et al. Recovery of zinc and manganese from spent alkaline batteries by liquid-liquid extraction with Cyanex 272. Journal of Power Sources (2003). , 115-367.

[9] Bernardes, A. M, Espinosa, D. C. R, & Tenório, J. A. S. Recycling of batteries: a review of current processes and technologies. Journal of Powder Sources (2004). , 130-291.

[10] Nougueira, C. A, & Margarido, F. Battery recycling by hydrometallurgy: evaluation of simultaneous treatment of several cell systems. Energy Technology (2012). , 227-234.

[11] Komilis, D, Bandi, D, Kakaronis, G, & Zouppouris, G. The influence of spent house-hold batteries to the organic fraction of municipal solid wastes during composting. Science of the Total Environment. (2011)., 409-2555.

[12] Brenniman, G. R, et al. Automotive and household batteries. In: KREITH, Frank, Handbook of Solid Waste Management, New York: McGraw-Hill. (1994)., 9.

[13] Rydh, C. J, & Svard, B. Impact on global metal flows arising from the use of portable rechargeable batteries. The Science of the Total Environment (2003)., 302-167.

[14] Salgado, A. L, Veloso, A. M. O, Pereira, D. D, Gontijo, G. S, Salum, A, & Mansur, M. B. Recovery of zinc and manganese from spent alkaline batteries by liquid-liquid ex-traction with Cyanex 272. Journal of Power Sources (2003)., 115-367.

[15] Sayilgan, E, Kukrer, T, Civelekoglu, G, Ferella, F, Akcil, A, Veglio, F, & Kitis, M. A review of technologies for the recovery of metals from spent alkaline and zinc-carbon batteries. Hydrometallurgy (2009)., 97-158.

[16] http://wwwmma.gov.br/port/conama/legiabre.cfm?codlegi=589 (Accessed 30 No-vember (2012).

[17] Provazi, K, Campos, B. A, Espinosa, D. C. R, & Tenório, J. A. S. Metal separation from mixed types of batteries using selective precipitation and liquid-liquid extrac-tion techniques. Waste Management (2011)., 31-59.

[18] Nan, J, et al. Recycling Spent Zinc Manganese Dioxide Batteries Through Synthesiz-ing Zn-Mn Ferrite Magnetic Materials. Journal of Hazardous Materials. (2006)., 133, 257-261.

[19] Rascio, D. C. et. al. Reaproveitamento de Óxidos de Manganês de Pilhas Descartadas para Eletrocatálise da Reação de Redução de Oxigênio em Meio Básico, Química No-va (2010)., 33(3), 730-733.

[20] Veloso, L. R. S, et al. Development of a hydrometallurgical route for the recovery of zinc and manganese from spent alkaline batteries, Journal of Power Sources (2005)., 152-295.

[21] Afonso, J. C. Processamento da Pasta Eletrolítica de Pilhas Usadas, Química Nova (2003)., 26(4), 573-577.

[22] Bocchi, N, Ferracin, L. C, & Biaggio, S. R. Pilhas e Baterias: Funcionamento e Impacto Ambiental. Química Nova na Escola (2000)., 11-3.

[23] Gervais, C, & Ouki, S. K. Effects of foundry dusts on the mechanical, microstructural and leaching characteristics of a cementitious system. Waste Management Series (2000)., 1-782.

[24] Belardia, G, Lavecchiab, R, Medicib, F, & Piga, L. Thermal treatment for recovery of manganese and zinc from zinc-carbon and alkaline spent batteries. Waste Manage-ment, (2012)., 32(10), 1945-1951.

[25] Rossi, G. Biohydrometallurgy. McGraw-Hill, Hamburg (1990).

[26] Rohwerder, T, Gehrke, T, Kinzler, K, & Sand, W. Bioleaching review part A: Progress in Bioleaching, fundamentals and mechanisms of bacterial metal sulfide oxidation. Applied Microbiology Biotechnology. (2003). , 63-239.

[27] Xin, B, Jiang, W, Aslam, H, Zhang, K, Liu, C, Wang, R, & Yutao, W. Bioleaching of zinc and manganese from spent Zn-Mn batteries and mechanism exploration. Bioresource technology (2012). , 106-147.

[28] da Silva PS., Schmitz E. P. S., Spinelli A., Garcia J. R. Electrodeposition of Zn and Zn-Mn alloy coatings from an electrolytic bath prepared by recovery of exhausted zinc-carbon batteries. Journal of Power Sources (2012). , 210-116.

[29] Brito, P. S. D, Patrício, S, Rodrigues, L. F, & Sequeira, C. A. C. Electrodeposition of Zn-Mn alloys from recycling Zn-MnO$_2$ batteries solutions. Surface & Coatings Technology (2012). , 206-3036.

[30] Ballesteros, J. C, Díaz-arista, P, Meas, Y, Ortega, R, & Trejo, G. Zinc electrodeposition in the presence of polyethylene glycol 20000 Electrochimica Acta (2007). , 52-3686.

[31] Díaz-arista, P, Ortiz, Z. I, Ruiz, H, Ortega, R, Meas, Y, & Trejo, G. Electrodeposition and characterization of Zn-Mn alloy coatings obtained from a chloride-based acidic bath containing ammonium thiocyanate as an additive. Surface & Coating. Technology (2009). , 203-1167.

[32] Sylla, D, Creus, J, Savall, C, Roggy, O, & Gadouleau, M. Refait Ph. Electrodeposition of Zn-Mn alloys on steel from acidic Zn-Mn chloride solutions. Thin Solid Films (2003).

[33] Savall, C, Rebere, C, Sylla, D, & Gadouleau, M. Refait Ph., Creus J. Morphological and structural characterisation of electrodeposited Zn-Mn alloys from acidic chloride bath. Material Science & Engeneering (2006). A 430 165.

[34] Bucko, M, Rogan, J, Stevanovi, S. I, Peri-gruji, A, & Bajat, J. B. Initial corrosion protection of Zn-Mn alloys electrodeposited from alkaline solution. Corrosion Science (2011). , 53-2861.

[35] Ortiz, Z. I, Díaz-arista, P, Meas, Y, Ortega-borges, R, & Trejo, G. Characterization of the corrosion products of electrodeposited Zn, Zn-Co and Zn-Mn alloys coatings. Corrosion Science (2009). , 51-2703.

[36] Gomes, A. da Silva Pereira M.I. Pulsed electrodeposition of Zn in the presence of surfactants. Electrochimica Acta (2006). , 51-1342.

[37] Song, K. D, Kim, K. B, Han, S. H, & Lee, H. Effect of additives on hydrogen evolution and absorption during Zn electrodeposition investigated by EQCM. Electrochemistry & Solid-State Letter (2004). CC24., 20.

[38] Díaz-arista, P, Antano-lópez, R, Meas, Y, Ortega, R, Chainet, E, Ozil, P, & Trejo, G. EQCM study of the electrodeposition of manganese in the presence of ammonium thiocyanate in chloride-based acidic solutions. Electrochimica Acta (2006).

[39] Atlas of Eh-pH diagramsIntercomparison of Thermodynamic Geological Survey of Japan Open File Report National of Advanced Industrial Science and Technology, May (2005). www.gsj.jp/GDB/openfile/files/no0419/openfile419e.pdf.Accessed 2012. (419)

[40] Kim, J. W, Lee, J. Y, & Park, S. M. Effects of Organic Additives on Zinc Electrodeposition at Iron Electrodes Studied by EQCM and in Situ STM. Langmuir (2004). , 20-459.

Chapter 9

Surface Modification of Nanoparticles Used in Biomedical Applications

Evrim Umut

Additional information is available at the end of the chapter

1. Introduction

In the last two decades a lot of attention has been payed on the preparation of nanoscaled materials and recently, depending on the development of new fabrication and characterization techniques, materials composed of a few atoms up to hundreds of atoms can be synthesized and their properties determined easily. Nano sized materials, as compared to their bulk counterparts, exhibit new characteristic optical, electrical and magnetic properties due to the enhanced surface to volume ratio and quantum confinement effects emerging in these size ranges. These new features of nanoparticles offers them the possibility to be used in a wide range of technological (magnetic data storage, refrigeration), environmental (catalysts, hydrogen storage), energy (lithium-ion batteries, solar cells) and biomedical applications. In biomedical applications there are different kind of nanoparticles used like metallic [1], magnetic, fluorescent (quantum dot) [2,3], polymeric [4,5] and protein-based nanoparticles [6, 7], in which much of the research in this field is focused on the magnetic nanoparticles. In this review only magnetic nanoparticles, which are composed of a magnetic core surrounded by a functionalized biocompatible surface shell will be concerned, where several reviews on other type of nanoparticles are available in the literature. In the scope of the text particular attention will be payed on superparamagnetic iron oxide (SPIONs), which is beyond the most studied one among all types of magnetic nanoparticles. In the beginning of the article, the biomedical applications of magnetic nanoparticles are summarized together with the key factors effecting the nanoparticles' performance in these applications. Then the requirement for the surface treatment of the nanoparticles are discussed in the context of colloidal stability, toxicity (biocompatibility) and functionalization. Finally, the followed procedures for the surface coating of magnetic nanoparticles are briefly explained and the different materials used as surface coatings are listed in detail with examples from literature.

2. Biomedical applications of magnetic nanoparticles

Magnetic nanoparticles (MNP) with dimensions ranging from a few nanometers up to tens of nanometers, thanks to their comparable or smaller size than proteins, cells or viruses, are able to interact with (bind to or penetrate into) biological entities of interest [8]. These size advantages of MNPs together with their sensing, moving and heating capabilites based on the unique nanometer-scale magnetic and physiological properties give them the possibility to be used in biomedical applications such as magnetic resonance imaging (MRI), targeted drug delivery and hyperthermia [9].

In MRI, which is the most promising non-invasive technique for the diagnosis of diseases, MNPs are used as contrast enhancement agents [10-12]. The improved contrast in MR images permits better definition and precise locating of diseased tissues (such as tumors) together with monitoring the effect of applied therapy. The operation of MRI is based on the Nuclear Magnetic Resonance (NMR) phenomena and the image processing is realized by spatially encoding of NMR signal of water protons which comes from different volume elements in the body called voxels. The image contrast in MRI depends mainly on proton density, spin-lattice (T_1) and spin-spin (T_2) nuclear relaxation times, differently weighted along different parts (voxels) of the body. The so-called contrast agents (CAs) themselves do not generate any signals, yet they contribute to the nuclear relaxation of water protons by creating local magnetic fields, which are fluctuating in time through different mechanisms like magnetization reversal and water diffusion [13]. As a consequence, the CAs decrease or increase the MRI signal intensity in the tissues by shortening both the T_1 and T_2 relaxation times of nearby protons resulting darker or brighter points in the image. The contrast enhancement efficiency of CAs is measured by the relaxivity $r_{1,2}$, which is defined as the increament of the nuclear relaxation rates $1/T_{1,2}$ of water protons induced by one mM of the magnetic ion. The CAs having a ratio r_2/r_1 greater than two, especially at magnetic fields mostly used in MRI tomography (0.5, 1,5 or 3 Tesla), are classified as T_2-relaxing (or negative) CAs since they more effectively decrease T_2 rather than T_1. On the other side CAs, characterized with a ratio r_2/r_1 smaller than two, have more pronounced effect on T_1 and hence called as T_1-relaxing (or positive) contrast agents [14]. The MNPs showing superparamagnetic property at physiological temperatures generally serve as T_2-relaxing CAs and they negatively improve the image contrast resulting darker spots where they are delivered. Commercially a wide variety of superparamagnetic iron oxide (SPIO) based negative CAs are available in the market like Endorem, Sinerem, Resovist, Supravist, Clariscan, Abdoscan etc., where each of them are used for different puposes or in different organs in clinical MRI application.

In a second biomedical application called magnetic hyperthermia, which is a thermally treatment of cancerous cells based on the fact that the cancer cells are more susceptible to high temperatures than the healthy ones, the MNPs can be used as heating mediators [15,16]. In this technique after concentrating the MNPs in the region of malignant tissue (by targeting or by direct injection), the MNPs are made to resonantly respond to a time-varying magnetic field and transfer energy from the exciting field to the surroundings as heat. By this way using an alternating field with sufficient intensity and optimum frequency, the temperature of tissue can be increased above 40-42 °C and the infected cells could be selectively destroyed. According to the models describing the heat release mechanisms of MNPs, increasing the frequency and the amplitude of the alternating field promises to significantly enhance the amount of released heat, but the limitations imposed by the biological systems restrict these values under a few tens of kA/m and a few hundreds of kHz for the field strength and frequency, respectively [17]. The MNPs' heating capacity in magnetic hyperthermia is denoted by Specific Absorbtion Rate (SAR) or in another term Specific Loss of Power (SLP), which is a measure of the energy converted into heat per unit mass. As the similar case in MRI CAs, the majority of magnetic hyperthermia heat mediators investigated to date are based on iron oxide MNPs, where in these studies the typical values reported for the maximum attained SAR range between 10 and 200 W/g [18-20]. However, it has been also focused on several systems alternative to iron oxide, where in some of them SAR values 3-5 times larger than those of similar iron oxide MNPs are attained for the same field parameters [21-22]. Recently, although there are lots of in-vitro studies about magnetic hyperthermia, the therapy with hyperthermia is still in pre-clinical stage and only a few studies on human patients are reported [23,24]. However, in cancer treatment the magnetic hyperthermia is thought to be introduced as a complementary technique to chemo- and radiotherapy as increasing the effects of these therapies [25, 26].

In another major in-vivo application, the MNPs are used as drug carriers in a magnetic 'tag-drag-release' process called targeted drug delivery. In a drug delivery process the MNPs, loaded with special drug molecules or conventional chemotherapy agents, are directly vectorized to tumor cells by targeting ligands on their surfaces or they brought into the vicinity of target tissue through magnetic forces exerted on them under an applied external magnetic field. Once the drugs/carriers are concentrated at the diseased site, the drugs are released from the carriers, again through modulation of magnetic field, enzymatic activity or changes in physiological conditions such as pH, osmolality or temperature. With this approach, the tumor cells can be destroyed by concentrating only the required quantity (dose) of drugs at target specific locations with minimized side effects on healthy tissues. The performance of the application depends mainly on the drug release kinetics and the cellular uptake of MNPs in tissues. There are huge number of drug delivery studies in the literature reporting both in-vitro and in-vivo results on different cell cultures and different types of tumors respectively, where in these studies several kind of targeting ligands and anticancer drugs are tested and in most of them again superparamagnetic iron oxide is used as magnetic core [27,28]. Generally in the design of MNPs for targeted drug delivery, in order to monitor the effect of the therapy, MRI contrast increament ability of the same MNP system is also investigated [29].

Actually in recent years much more interest has been concentrated on multifunctional MNPs, in which the above mentioned diagnostic (MRI) and therapeutic (hyperthemia and drug delivery) capabilities are combined [15,30,31]. Although some MNP-based MRI contrast agents are commercialy available on one side and magnetic hyperthermia is already utilized in conjunction with other kind of therapies on the other side, MNPs optimized to perform both functions (diagnostic and therapeutical) have not been developed yet. Indeed the possibility to associate therapeutic effect generated by the heat release and delivered drugs with the enhanced contrast in MRI images, is extremely appealing since it would provide the possibility before heating the tissue to track the particle distribution by MRI, and after the drug therapy or thermotherapy to have an immediate control of the efficacy of the treatment itself. In Figure.1 a summarized illustration of biomedical applications has been shown.

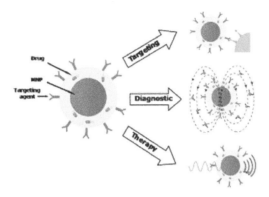

Figure 1. Biomedical applications of magnetic nanoparticles (image has been reproduced from A. Lasicalfari et al. [9].

3. Design of MNPs for biomedical applications

In many biomedical applications of MNPs, usually a core/shell structure is required, where the inorganic magnetic core is surrounded by an outer layer of shell (coating). The successfull design of MNPs needs a carefull selection of magnetic core and surface coating material, where the first mainly determines the MNPs' above mentioned heating, sensing etc. abilities related with application eficiency and the second specifies the interaction of these MNPs with physiological environment. In following sections the properties of magnetic core and the surface coating are discussed in detail.

3.1. Magnetic core

In the selection of magnetic core some important aspects should be taken in the consideration, such that for the first the magnetic core should be crystalline and smaller than a critical size as to consist of only one magnetic domain. This ensures that the single-domain nanoparticles exihibit superparamagnetic behaviour with zero remanent magnetization in the absence and a very high magnetization (approximately three orders of magnitude greater than paramagnetic materials) in the presence of an external field. This physical phenomenon is the key requirement for biomedical applications since the particles can be dispersed and concentrated in solution in-vitro or in blood circulation in-vivo, without forming magnetized clusters and they can also respond to an instantly applied field with some kind of magnetic on/off switching behaviour. For the second aspect, the size distribution of the magnetic cores should be as narrow as possible and third, all the magnetic cores in a particular sample should have a unique and uniform shape (monodispersed). This is because all the magnetic and physico-chemical properties strongly depend on the size and shape of the magnetic cores. From an applicative point of view the size, properly speaking the hydrodynamic size which is the total diameter of MNP including the coating tickness, is also important for the elongation of MNPs' circulation time in blood and for the improvement of their internalization by the cells at the target tissue such that smaller nanoparticles have bigger chance to reach the target cells and to penetrate inside them.

In the search of suitable elements for the magnetic core of MNPs, among other magnetic materials transition metals like Fe, Ni, Co and Mn are good candidates since they offer high magnetization values which is important for high performance MRI and hyperthermia applications. However they are not stable and oxidate very quickly yet in the synthesis step, if they are not specially treated. For this reason transition metal oxide compounds (also called ferrites) which are stable and have acceptable magnetizations are generally introduced in biomedical applications. Superparamagnetic iron oxide (SPION), belonging to ferrite family, is the most commonly employed one in biomedical applications. Nano-crystalline iron oxides have an inverse spinel crystal structures, where the oxygen atoms form face centered cubic lattices and iron ions occupy tetrahedral (T_d) and octahedral (O_h) interstitial sites (Figure.2). İron oxide generally exist as two stable forms called magnetite (Fe_3O_4) and its γ phase maghemite (γ-Fe_2O_3), but there is also a α phase called hematite (α-Fe_2O_3), which is not stable and obtained by thermal treatment of magnetite or maghemite. In magnetite, bivalent Fe^{+2} ions occupy O_h sites and trivalent Fe^{+3} ions are equally distributed between O_h and T_d sites, whereas maghemite, which can be result from the oxidation of magnetite, only contains Fe^{+2} ions distributed randomly over O_h and T_d sites [32]. In magnetite, since there is the same number of Fe^{+3} ions in O_h and T_d sites, which compensate for each other, the resulting magnetization arises only from the uncompensated Fe^{+2} ions in O_h sites. On the other side the magnetization of maghemite originates from uncompensated Fe^{+3} ions. However their magnetic behaivours and other properties are quite similar, which makes it very difficult to distinguish between them.

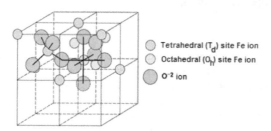

Figure 2. The cubic inverse spinel crystal structure of iron oxide showing T_d and O_h sites

Alternatively other types of ferrites were also studied for biomedical applications. In these ferrites, as compared to the iron oxide nanocrystalls, Fe^{+2} ions are fully or partially replaced by other transition metals in spinel structure and they represented by a general formula

(MFe_2O_4; M=Zn, Ni, Co, Mn). Manganese ferrites ($MnFe_2O_4$) serve as potential MRI contrast agents with largest magnetization among other ferrites [33] and zinc ferrite ($ZnFe_2O_4$) nanocrystals demonstrated better MRI contrast with respect to similar magnetite nanocrystals [34]. On the other side as regard to magnetic hyperthermia, the use of cobalt ferrites ($CoFe_2O_4$), known by its high magnetic anisotropy energy which is responsible for holding the magnetization along certain direction, has proven to be a good way since much higher heating rates were reported for these nanocrystals compared to other ferrites [21]. Another strategy followed in the design of magnetic core for biomedical applications is the synhesis of mixed ferrites, where simple ferrites including one kind of magnetic ion except iron are doped with other kind of magnetic ion. This is generally realized in order to utilize from different outstanding magnetic features of different ions. For example in hyperthermia application, Co, being a hard magnetic material, is doped to other ferrites (MFe_2O_4) in changing concentrations ($Co_xM_{1-x}Fe_2O_4$; x=concentration) in order to increase the magnetic anisotropy eventually to improve the heat transfer rate, whereas Zn is added for reducing the Curie temperature of resulting mixed ferrite [35]. This latter operation permits the tuning of the maximum reached temperature by heat transfer and prohibits overheating of healthy tissues via the process called self-controlled hyperthermia [36]. Table.1 summarizes some important magnetic parameters of transition metal oxides (ferrites) used in biomedical applications.

Another class of materials used as magnetic cores are the magnetic alloy nanoparticles, which composed of two or three different kind of metals like FeCo, FePt and NiCu. FePt is the most famous one among these materials due to its chemical stability and high magnetic anisotropy. Maenosono and Saita have studied FePt nanocrystals and proposed them to be used as high performance contrast agents and heating mediators in MRI and magnetic hyperthermia, respectively [37,38].

Ferrites	RT saturation magnetization M_s (emu/g)	Anisotropy constant K_1 (x10^4 J/m^3)	Curie temperature T_c (°C)	Superparamagnetic size D_{SP} (nm)
Fe_3O_4	90-100	-1.2	585	25
$NiFe_2O_4$	56	-0.68	585	28
$CoFe_2O_4$	80-94	18-39	520	14
$MnFe_2O_4$	80	-0.25	300	25

Table 1. Some important magnetic parameters (room temperature saturation magnetization, first degree anisotropy constant, Curie temperature and superparamagnetic transition size) of ferrites [32]

Although MNPs can be prepared with several techniques including inert gas condensation [39], mechanical milling [40], spray pyrolysis [41], sol-gel [42], vapor deposition [43] and wet chemical processes [44], all the before mentioned key requirements for biomedical applications like crystallinity, size and shape uniformity together with other criterias for organic/inorganic coating of MNP, can be satisfied by hydrothermal chemical decomposition method, which is the most common one used in the synthesis of MNPs for biomedical applications. In this liquid phase synthesis method, basically several organometalic precursors with suitable sitochio-metric ratios are put in reaction in the presence of some organic surfactants or polimers. During the decomposition process at high temperature, in order to prohibit a possible oxidation, continously an inert gas is fluxed through the mixture and at the end of the reaction the desired nanomaterial is obtained as a precipitate. The method enables the control on size of the MNPs through systematically change of the reaction parameters like concentration, reaction temperature and reaction time. In the method, MNPs with very narrow size distributions (σ~% 10) are synhesized and the distribution can be further improved (down to σ~% 5) by subsequently precipitating, redispersing and centrifuging of the particles, so-called as size selection processes (Figure.3). S. Sun and co-workers have published several pioneering papers in this area, where they have introduced hydrothermal routes for the synthesis of size-controlled MNPs for the first time and they have synthesized monodispersed iron oxide [45], other transition metal oxides [46] and iron-platinum [47] nanoparticles by using this method. The other advantages of the methods are the abilities of large quantity production of MNPs and the subsequent coating of them which will be explained in detail in the following of the chapter.

Another common technique used in MNP synthesis for biomedical applications is the co-precipitation method. It has some disadvantages against thermal decomposition method like lower crystallinity and lower monodispersity but the procedure is easier and at the end of the procedure larger amount of product can be yield. The method is based on the simultaneous nucleation and growth of magnetic cores by dissolving metal salt precursors in aquaeous environment with changing pH and temperature [35]. Alternatively this reaction can be governed in a confined environment via microemulsion method. In this method the co-precipitation reaction takes place in some kind of "nano-reactors" called micelles, those are dispersions of two immiscible liquids like oil-in-water or water-in-oil (reverse micelle). The main advantage of the microemulsion is to ensure that the reaction occurs in an isolated media limiting the particle growth and it is possible to control the size of the magnetic core precisely by changing the size of the micelles [48].

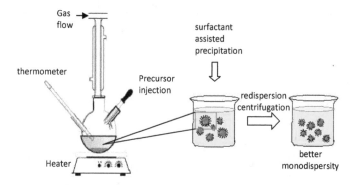

Figure 3. An illustration of nanoparticle synthesis by high temperature thermal decomposition method and size selection processes.

3.2. Surface coating

In the design of MNPs, except the selection of a suitable magnetic core, fine tuning of surface coating materials represents a major challenge for the practical use of MNPs in clinical applications. The coating can consist of long-chain organic ligans or inorganic/organic polymers, where these ligands or polymers can be introduced during (in-situ coating) or after (post-synthetic coating) synthesis. During the in-situ coating, which is the procedure followed in co-precipitation synthesis technique, precursors of magnetic cores and coating materials are dissolved in the same reaction solution, and the nucleation of magnetic core and the coating occurs simultaneously. On the contrary in post-synthetic coating, which is the case in MNP synthesis with thermal decomposition technique, the surface coating materials are introduced after the formation of magnetic cores. In both procedures in order to link the surface molecules to magnetic cores, generally two different approaches; either end-grafting or surface-encapsulation are followed. In the former one, the coating molecules are anchored on magnetic core by the help of a single capping group at their one end, whereas in the latter generally polymers, already carrying multiple active groups, are attached on the surface of magnetic core with multiple connections resulting stronger and more stable coatings.

In some cases, surface modification of MNPs can drammatically change the magnetic properties hence the performance of the MNPs in biomedical applications depending on which coating material is used and how these materials are linked on the magnetic core surface as discussed above. Formation of chemical bonds between coating molecules and surface metal ions changes the surface spin structure and consequently the magnetic properties of coated MNPs with recpect to uncoated ones. Actually it is difficult to discriminate between surface coating and finite size contributions, the latter being the effect of surface spin canting due to the minimization of magnetostatic energy at the surface and observed also in uncoated particles [49]. In finite size effect the canted spins at MNP's surface, do not respond to an external applied magnetic field as the bulk spins and give rise to a significant decrease in net magnetization, where this effect becomes dominant in smaller MNPs, since the volume fraction of disordered surface spins are increased [50]. On the other side, the physical origin of the surface coating effect on magnetic properties is still unclear and different results were reported in several studies for different kind of magnetic cores and coating materials. For example, Vestal and Zhang [51] have investigated the influence of the surface coordination chemistry on the magnetic properties of $MnFe_2O_4$ nanoparticles by capping the 4, 12 and 25 nm sized $MnFe_2O_4$ nanoparticles with a variety of substituted benzenes and substituted benzoic acid ligands and observed an increase in saturation magnetization, whereas a decrease is reported by Ngo et al. [52] for citrate coated 3 nm $CoFe_2O_4$ nanoparticles. On the other hand, no significant change in the magnetic properties of 10 nm sized γ-Fe_2O_3 nanoparticles is observed for different surface chemical treatments such as NO_3, ClO_4 and SO_4 [53]. Another in-direct effect of the surface coating on magnetic properties, which should be mentioned, is to reduce the magnetic interparticle interactions through decreasing the distance between magnetic cores. This is important since strong interparticle interactions can alter the MNPs' overall magnetic properties i.e. superparamagnetic behaivour and diverse it from isolated ones.

Dormann and Fiorani have published several important papers investigating the magnetic interparticle interactions and the effect of coating [54-56].

The surface coating of MNPs plays a crucial role in biomedical applications by fulfilling more than one function at a time. The organic/inorganic surface coating is important for i) prohibiting agglomeration (clustering) of MNPs due to the above mentioned interparticle interactions and eventually providing the colloidal stability of water/organic solvent based suspensions/ solutions (ferrofluids) prepared with MNPs ii) providing biocompatibility of MNPs by preventing any toxic ion leakage from magnetic core into the biological environment iii) serving as a base for further anchoring of functional groups such as biomarkers, antibodies, peptides etc. In the next paragraphs these functions of MNP's surface coating are discussed in detail.

3.2.1. Stability

The stabilization of the MNPs is crucial to obtain magnetic colloidal ferrofluids that are stable against aggregation both in a biological medium and in a magnetic field. By coating the MNPs, direct contacts among the particles are prevented by surface ligands and polymer chains extending into the medium. Therefore, no aggregates of particles will be formed or the rate of precipitation will be decreased depending on the degree of covarege of coating on the particles and the thickness of the coating layer. The stability of a magnetic colloidal suspension results from the equilibrium between attractive (magnetic dipol-dipol, van der Waals) and repulsive (electrostatic, steric) forces [57]. Controlling the strength of these forces is a key parameter to elaborate MNPs with good stability. A suitable surface coating should not only keep the MNPs apart from each other, eliminating the distance dependent attractive forces but should also ensure the charge neutrality and steric stability of MNPs in aquaeous or organic media. The MNPs capped by hydrophilic surfactants like dodecylamine, sodium oleate can be easily dispersed in aqua but when they are stabilized by hydrophobic surfactants like oleic acid, oleylamine, which is the frequent case in the MNP synthesis with thermal decomposition method, they can only be dispersed in nonpolar solvents such as hexane, toluene or weak polar solvents such as chloroform. Therefore in order to stabilize these MNPs in aquaeous media, different kind of polymers can be introduced on MNPs either by end-grafting or surface encapsulation or as an alternative strategy, the MNPs can be capped with amphiphilic molecules, consisting both hydrophobic and hydrophilic regions at their opposite ends via hydrophobic interactions resulting micelle-like structures [58].

3.2.2. Toxicity (Biocompatibility)

Another critical point which should be considered in the design of MNPs for biomedical applications is the toxicity of magnetic ion included in the magnetic core. The surface coating ensures a double- sided isolation both preventing the release of toxic ions from magnetic core into biological media and shielding the magnetic core from oxidization and corrosion. Among different types of MNPs iron oxide is by far the most commonly employed one for in vivo applications since iron is physiologically well tolerated. This is partially because the body is designed to process excess iron and it is already stored primarily in the core of the iron storage protein called ferritin. Manganese (Mn) and Zinc (Zn) are other essential trace elements in human bodies, but their tolerable limit is much lower than iron's, so a surface manipulation is required [59]. Other elements, such as Co and Ni, which are desired for their high perform-ance in hyperthermia application, are highly toxic and necessitates proper coatings when they are used in vivo. During in vivo applications once the MNPs are injected into the body, they are recognized by the body's major defence system (also known as reticuloendothelial system (RES)), which eliminates any foreign substance from blood stream. In this "opsonisation" process, MNPs are attacked by plasma proteins, which are sent by RES and are responsible for the clearance of MNPs. The specific surface coatings prevent the adsorbtion of these proteins, thus elongate the MNPs circulation time in blood and maximize the possibility to reach target tissues [60].

3.2.3. Functionalization

In order to enable the direct use of MNPs in biomedical applications, the MNPs should be further functionalized by conjugating them with functional groups. The surface coating provides suitable base for the attachment of these functional groups on MNPs. These groups such as antibodies, peptides, polysaccharides etc. permit specific recognition of cell types and target the nanoparticles to a specific tissue or cell type by binding to a cell surface receptor. Usually some linker molecules such as 1-ethyl-3-(3-dimethylaminopropyl) carbodi-mide hydrochloride (EDCI), N-succinimidyl 3-(2-pyridyldithio) propionate (SPDP), N-hydroxysuc-cinimide or N, N'-methylene bis acrylamide (MBA) are also used to attach the initial hydro-philic coated molecules to these targeting units [58]. In practice although the targeted cell population is recognized with high specifity, the fraction of targeted cells interacting with the antibody attached MNP is relatively low. Actually the effectiveness of biomedical applications depends more on cell-nanoparticle interactions than particle targeting. It is indicated that the cell membranes plays an important role in cell-nanoparticle interactions, either particle attachment to the cell membrane or particle uptake into the cell body. In order to clarify the nature of cell-nanoparticle interactions, usually phospholipid bilayers mimicking a cellular membrane or living cell membranes are used in studies.

3.2.4. Surface coating materials

For utilizing all the above mentioned functionalities of MNP's surface coating, different criterias of inorganic/organic molecules and polymers such as hydrophobic or hydrophilic, neutral or charged, synthetic or natural etc. are considered. The polymer coatings can be classified as natural polymers such as chitosan, dextran, rhamnose or synthetic polymers like polyethyleneglycol (PEG), polyvinyl alcohol (PVA), polyethyleneimine (PEI), polyvinylpyro-lidone (PVP). In some cases other organic molecules including oleic acid, oleylamine, dode-cylamine and sodium oleate are also used to enhance water solubility of MNPs. On the other side, there are very less number of inorganic materials available for MNPs surface coatings, in which gold (Au) and silica (SiO_2) are the most common used ones due to their biocompatibility. In Figure.4 a representative sketch of MNPs with different types of coatings are shown. In the next paragraphs, these coating materials are introduced in detail by summarizing their outstanding properties and by discussing their feasibility in clinical applications with recent examples from literature.

Figure 4. A representative sketch of MNPs with different types of surface coatings: (a) inorganic materials (b) long chain organic molecules (c) organic polymers

3.2.4.1. Inorganic surface coatings

Gold (Au)

Nanosized gold (Au) attracts too much attention due to its unique physical properties combined with chemical stability, biocompatibility and surface properties which permits its attachment to different chemical moieties. Surface modification of MNPs with biocompatible gold, both provides the in-water stabilization of particles by preventing their agglomeration and enables their fuctionalization by the attacment of several ligands on them. From applicative point of view, it also maintains the incorporation of some optical properties on MNPs promoting their use in dual-mode (optical and magnetic diagnosis) applications. Gold nanosurface shows surface plasmon resonance (SPR) phenomena, which triggers MNPs to strongly absorb and scatter near infrared wavelengths (Surface Enhanced Raman Scattering-SERS) accepted as "clear window" for deeper penetration of light into the human tissue [61]. In this respect gold modified MNPs, besides other applications, can simultaneously be used in SERS imaging, where detection in molecular level is possible enabling the diagnosis of diseases at very early stages. E. Umut et al. [12] have synthesized organically coated mono-dispersed gold-iron oxide hybrid nanoparticles following the chemical procedures introduced by W. Shi [62] and H. Yu [63], and showed that the superparamagnetic hybridnanoparticles have MRI contrast enhancement abilities associated with optical properties based on SPR phenomena, where the optical SPR absorbance depens on the geometry of synthesized hybrid nanoparticles. Similarly X. Ji et al. [64] have fabricated hybridnanoparticles where superparamagnetic iron oxide nanoparticles were embedded in silica and further coated with gold nanoshell, and they exihibit that the resulting nanoparticles both can improve the MRI contrast and have efficient photothermal effects when exposed to near infrared light. In another study H. Y. Park et al. [65] have synthesized Fe_3O_4@Au coreshell nanoparticles with controllable size ranging from 5 to 100 nm following a hetero-interparticle coalescence strategy and demonstrated the surface protein-binding properties and SERS effectiveness of synthesized nanoparticles. There are lots of similar studies in the literature, where gold nanoparticles or nanosurfaces are introduced in different "smart" nanostructures for utilizing its optical properties together with surface binding properties in multifunctional manner [66-69].

Silica (SiO$_2$)

Another inorganic but "biofriendly" material used as MNPs' coating is silica, which is known for its chemical stability and easy-to-formation. The first advantage of having surface enriched in silica is the presence of silanol groups, which can easily react with coupling agents providing strong attachment of surface ligands on MNPs [70]. As a second advantage, silica coating increases the stabilization of MNPs in liquid dispersions both by preventing the dipolar attractions and by increasing the surface charges hence the electrostatic repulsions between particles in non-aquaeous dispersions. There are several succesfull methods available for the formation of silica coating, in which the mostly used one is the Stöber method, where a hydrolysis reaction of tetraethyl ortosilicate (TEOS) is governed in alcohol media under catalysis by ammonia [71]. In different studies reporting the synthesis of silica coated iron oxide nanoparticles, Y. H. Deng et al. have used Stöber method and obtained spherical coreshell nanoparticles by the condensation of TEOS in sol-gel form on pre-formed magnetite nanoparticles and have investigated the morphology and the thickness of the coating by systematically changing the used alcohol type or the amount of alcohol, amonia and TEOS [72]. S. Santra et al. have applied water-in-oil microemulsion method again for the coating of previously synthesized iron oxide nanoparticles by using different nonionic surfactants and obtained as small as 1-2 nm and very uniformly sized (with standard deviation less than 10%) nanoparticles [73]. In another study D. K. Yi and colleagues, again by following a reverse microemulsion method, have synthesized homogenous silica-coated SiO$_2$/Fe$_2$O$_3$ nanoparticles with changing shell thickness at first stage and then they used these nanoparticles to derive mesoporous silica-coated SiO$_2$/Fe$_2$O$_3$ MNPs and hollow SiO$_2$ nanoballs [74]. In these studies the effect of coating and coating tickness on the magnetic properties of iron oxide were also investigated. As being in Au coated MNPs, attachment of different functional components to silica encapsulated MNPs either by embedding into or binding onto silica shell, smart engineering nanostructures realizing more than one function can be prepared [75,76].

3.2.4.2. Organic surface coatings

Chitosan

Natural polymers and their derivatives have been widely utilized for coating of MNPs for in vivo applications. This is because such polymers are inexpensive and are known to be nonimmunogenic and nonantigenic in the body. They are usually anchored onto the surfaces of MNPs through carboxylate groups on their side chains. Chitosan is a natural polysaccharide cationic polymer, which is nontoxic, hydrophilic and biodegredable. There are lots of studies introducing chitosan as MNP carriers with the aim of use in biomedi-cal applications. Y. Chang and D. Chen [77] have reported the preparation of 13,5 nm sized chitosan coated magnetite nanoparticles, where the chitosan was first carboxymethylated and then covalently bound on the surface of preformed magnetite nanoparticles via carbodiimide activation. They have also demonstrated that the synhesized nanoparticles can be efficiently used in magnetic ion separation. Thereafter instead of this two step suspension crosslinking approach, J. Zhi et al. proposed an alternative method for in situ preparation of chitosan-magnetite nanocomposites in water-in-oil microemulsion [78]. By this method they have synthesized spherical chitosan nanoparticles in varying size from 10 nm to 80 nm with cubic shaped magnetite core. D. Kim et al. have prepared chitosan and -its derivative- starch coated iron oxide nanoparticles with the aim of treatment carcinoma cells by magnetic hyperthermia [79]. After performing in vitro cell cytotoxity and affinity tests together with magneto-caloric measurements on magnetic fluids, they have shown that targeting of MNPs to cells was improved by using a chitosan coating and the coated MNPs are expected to be promising materials for use in magnetic hyperther-mia. In another study published by Y. Ge et al. chitosan coated maghemite nanoparticles were modified with fluorescent dye by covalent bonding for dual-mode high efficient cellular imaging. They have shown that prepared nanoparticles could be efficiently internalized into cancer cells and serves as MRI contrast agents and optical probes for intravital fluorescence microscopy [80].

Dextran

Another natural polymer is dextran, which is a neutral, branched polysaccharide composed of glucose subunits. Dextran is one of the most frequently chosen polymer because of its high biocompatibility. The studies conducted in order to find the biocompatibility of dextran have shown that the surface-immobilized dextran on MNPs is stable in most tissue environment, because dextran is resistant to enzymatic degradation [81]. However, the cellular uptake of dextran coated MNPs are not sufficient for most biomedical applications. One strategy to overcome this handicap has been reported by A. Jordan et al. as creating aminosilane groups on the surface of MNPs and by following this approach in vitro cellular uptake of such nanoparticles in carcinoma and glioblastoma cells was found to be thousand times higher than that of only dextran-coated MNPs [82]. With this method also much more efficient hyperthe-mia results have been obtained. Another pathway to increase the internalization of dextran coated particles by tumor cells is the further coupling of them with specific ligands. Some authors, with the aim of achieving better localized MRI contrast, have attached "transferrin" onto dextran coated MNPs and the cellular uptake of these ligand modified MNPs were two or four times higher compared to unmodified ones [83, 84]. There are many studies reporting different synthesis methods for preparing dextran-coated iron oxide nanoparticles [85,86] and investigating their interaction with cells [87,88].

Polyethylene Glycol (PEG)

PEG is a neutral, hydrophilic, linear synthetic polymer that can be prepared with a wide range of terminal functional groups. By varying these functional groups, PEG can be binded to different surfaces. PEG coated MNPs reveal excellent stability and solubility in aquaeous dispersions and in pysiological media. Moreover the implementation of PEG as a surface coating of MNPs dramatically increases the blood circulation time of MNPs, enhancing their hinderence from the body's defence system. But as compared to multi-branched dextran coating, surface immobilized PEG permits limited grafting of further macromolecules since it has only one site available for ligand coupling. Related with the cellular uptake performance, Zhang et el. [89] have shown that the amount of internalized PEG coated MNPs into mouse macrophage cells are much lower than uncoated MNPs. However, for breast cancer cells PEG modification of MNPs promotes better internalization of particles, where this situation is explained with the high solubility of PEG in physiological media hence the possibility of its diffusion into cell membranes [90]. In any case, as being in dextran coated MNPs, additional surface coupling of targeting ligands on PEG modified MNPs increases their cellular uptake. In order to attain a better intracellular hyperthermia efficacy, M. Suzuki et al. have attached monoclonal IgG antibody on the surface of PEG-modified magnetite nanoparticles and showed that cellular uptake of particles is improved [91].

Polymers / organic molecule	Properties	Advantages / Disadvantages
Chitosan	-natural, cationic, hydrophilic, linear, biodegredable	-can be used in non-viral gene delivery
Dextran	-natural, branched, hydrophilic, biocompatible	-permits the anchoring of biovectors and drugs when functionalized with amino groups
Polyethyleneglycol (PEG)	-synthetic, neutral, hydrophilic, linear, biocompatible	-remain stable at high ionic strengs of solutions with varying PH values, enhances blood circulation time (a few hours), permits functionalization
Polyethyleneimine (PEI)	-synthetic, cationic, linear or branched, non-biodegredable, toxic	-forms strong covalent bonds with MNP's surface, can be used for DNA and RNA delivery, but exhibit cytotoxity
Polyvinylealcohol (PVA)	-synthetic, hydrophilic, biocompatible	- irreversibly binds on MNP's surface but can be used in temperature sensitive heating or drug release applications due to its decomposition temperatures (40-50 °C),
Polyvinylpyrolidone (PVP)	-synthetic, branched, hydrophilic	-forms covalent bonds with drugs containing nucleophilic functional groups

Table 2. Some properties of different organic surface coating materials

Polyvinyl Alcohol (PVA)

PVA is a hydrophilic, synthetic polymer. Coating of MNPs surface with PVA enhances the colloidal stability of ferrofluids prepared with these MNPs. But it has been suggested that PVA irreversibly binds on MNPs surface due to interconnected network with interface, which means a fraction of PVA remains associated with the nanoparticles despite repeated washing [92]. The residual PVA, in turn, influence different properties of nanoparticles such as particle size, zeta potential and surface hydrophobicity. Importantly, nanoparticles with higher amount of residual PVA had relatively lower cellular uptake. It is proposed that the lower intracellular uptake of nanoparticles with higher amount of residual PVA could be related to the higher hydrophilicity of the nanoparticle surface [93]. A. P. Fink et al. have coated 9 nm sized iron oxide nanoparticles with unfunctionalized or carboxylate, amine or thiol functionalized PVA and observed that nanoparticles coated with PVA and carboxyl and thiol functionalized PVA were non-toxic to melanoma cells, whereas for the amine functionalized PVA nanoparticles, some cytotoxicity was observed particularly when the polymer concentrations were high [94].

Polyethyleneimine (PEI)

PEI is a cationic, synthetic polymer and exist either as linear or branched forms. Although PEI is toxic and non-biodegredable, it has long been used for gene delivery thanks to its ability to bind with DNA [95]. Since it is a cationic polymer it can further interact with a wide variety of negatively charged complexes. Recently F. M. Kievit et al. have developed a complex MNP system, which is made of a superparamagnetic iron oxide nanoparticle (NP), which enables magnetic resonance imaging, coated with a novel copolymer comprised of short chain polyethylenimine (PEI) and poly(ethylene glycol) (PEG) grafted to the natural polysaccharide, chitosan (CP), which allows efficient loading and protection of the nucleic acids [96]. In this study they have illustrated the function of each component by comparative experiments and proposed that the designed complex MNP system is a potential candidate for safe in vivo delivery of DNA for gene therapy.

As should be summarized, there is a wide variety of coating materials could be attached on MNPs' surfaces by following different coating procedures and each of these materials have different advantages and disadvantages in biomedical applications depending on their characteristic properties like hydrophilicity, neutrality and structure. Table.2 lists some properties and outstanding advantages / disadvantages of different organic surface coating materials.

4. Conclusions

In the last two decades, a lot of attention has been devoted to synthesis and characterization of functionalized iron or other transition metal oxide based MNPs, which have potential use in diagnosis and/or therapy in cancer treatment. These MNPs can act as contrast enhancement agents in diagnostic applications such as MRI and/or they can be used as carriers or localy heat releasers in therapeutic applications such as targeted drug delivery and magnetic hyperthermia, respectively. In the design of MNPs, the selection of the magnetic core and its surface modification with several organic/inorganic materials and polymers plays the major role effecting the performance of MNPs in these biomedical applications. Among different available magnetic ions, the correct selection of the magnetic core for MNPs requires careful and balanced consideration on material's properties such as chemical stability, toxicity and magnetization. The magnetic core further should be crystalline, small than a critical size and have a narrow size distribution where all these requirements together with proper in-situ or post synthetic surface coatings are satisfied by chemical methods like co-precipitation and thermal decomposition method. The surface coating is important for ensuring the biocompatibility, colloidal stability and functionalization of MNPs, where a wide variety of coating materials are available like organic molecules/ polymers such as chitosan, dextran, Polyethyleneglycol (PEG), Polyethyleneimine (PEI), Polyvinylalcohol (PVA) or inorganic materials like silica and gold. Although much progress has been made on the fabrication of MNPs with delicate structure and enhanced surface properties, in using these MNPs for in vivo applications, major challenges still present like degredation, clearence of MNPs in the body, particle-cell interactions and changing physiological conditions like pH, temperature, blood pressure etc. which makes difficult to predict the behaivour of MNPs in biological medium.

Author details

Evrim Umut*

Address all correspondence to: eumut@hacettepe.edu.tr

Hacettepe University, Department of Physics Engineering, Ankara, Turkey

References

[1] Liao, H., Nehl, C. L. and Hafner, J. H., Biomedical Applications of Plasmon Resonant Metal Nanoparticles, Future Nanomedicine, 2006, 1 (2), pp. 201-208

[2] Corr, S. A., Rakovich, Y. P. and Gunko, Y. K., Multifunctional Magnetic-fluorescent Nanocomposites for Biomedical Applications, Nanoscale Research Letters, 2008, 3 (2), pp. 87-104

[3] Chatterjee, D. K., Gnanasammandhan, M. K. and Zhang, Y., Small Upconverting Fluorescent Nanoparticles for Biomedical Applications, Small ,2010, 6 (24), pp.2781-2795

[4] Soppimath, K. S., Aminabhavi, T. M., Kulkarni, A. R., Rudzinski, W. E., Biodegradable Polymeric Nanoparticles as Drug Delivery Devices, Journal of Controlled Release, 2001, 70 (1-2), pp.1-20

[5] Kumari, A., Yadav, S. K. and Yadav, S. C., Biodegradable Polymeric Nanoparticles Based Drug Delivery Systems, Colloids and Surfaces B:Biointerfaces, 2010, 75 (1), pp. 1-18

[6] Kogan, M. J., Olmedo, I., Hosta, L., Guerrero, A. R., Cruz, L. J., Albericio, F., Peptides and Metallic Nanoparticles for Biomedical Applications, Future Medicine, 2007, 2 (3), pp. 287-306

[7] Hawkins, M. J., Soon-Shiong, P. and Desai, N., Protein Nanoparticles as Durg Carriers in Clinical Medicine, Advanced Drug Delivery Reviews, 2008, 60 (8), pp.876-885

[8]   Varadan, V. K., Chen, L. and Xie, J., Nanomedicine: Design and Applications of Magnetic Nanomaterials, Nanosensors and Nanosystems, John Wiley and Sons Pub., 2008

[9]   Lascialfari, A. and Sangregorio, C., Magnetic Nanoparticles in Biomedicine: Recent Advances Chemistry Today, 2011, 29 (2), pp.20-23

[10] Casula, M. F., Floris, P., Innocenti, C., Lascialfari, A., Marinone, M., Corti, M., Sperling, R. A., Parak, W. J., Sangregorio, C., Magnetic Resonance Contrast Agents Based on Iron Oxide Superparamagnetic Ferrofluids; Chemistry of Materials, 2010, 22 (5), pp.1739–1748

[11] Masotti, A., Pitta, A., Ortaggi, G., Corti, M., Innocenti, C., Lascialfari, A., Marinone, M., Marzola, A., Daducci, A., Sbarbati, A., Micotti, E., Orsini, F., Poletti, G., Sangregorio, C., Synthesis and Characterization of Polyethylenimine-based Iron Oxide Composites as Novel Contrast Agents for MRI; Magnetic Resonance Materials in Physics, 2009, 22 (2), pp.77–87

[12] Umut, E., Pineider, F., Arosio, P., Sangregorio, C., Corti, M., Tabak, F., Lascialfari, A., Ghigna, P., Magnetic, Optical and Relaxometric Properties of Organically Coated Gold-magnetite (Au-Fe$_3$O$_4$) Hybrid Nanoparticles for Potential use in Biomedical Applications, Journal of Magnetism and Magnetic Materials, 2012, 324, pp. 2373-2379

[13] Gossuin, Y., Gillis, P., Hocq, A., Vuong, Q. L. and Roch, A., Magnetic Resonance Relaxation Properties of Superparamagnetic Particles, WIREs Nanomedicine and Nanobiotechnology, 2009, 1, pp. 299-310

[14] Bridot, J. L., Faure, A. C., Laurent, S., Riviere, C., Billotey, C., Hiba, B., Janier, M., Josserand, V., Coll, J. L., Vander Elst, L., Muller, R., Roux, S., Perriat, P., Tillement, O., Hybrid Gadolinium Oxide Nanoparticles: Multimodal Contrast Agents for in Vivo Imaging, Journal of American Chemical Society, 2007, 129 (16), pp. 5076-5084

[15] Kim, D. H., Nikles, D. E., Johnson, D. T., Brazel, C. S., Heat Generation of Aqueously Dispersed CoFe$_2$O$_4$ Nanoparticles as Heating Agents for Magnetically Activated Drug Delivery and Hyperthermia, Journal of Magnetism and Magnetic Materials, 2008, 320 (19), pp.2390–2396

[16] Gonzales, M., and Krishnan, K. M., Synthesis of Magnetoliposomes with Monodisperse Iron oxide Nanocrystal Cores for Hyperthermia, Journal of Magnetism and Magnetic Materials, 2005, 293 (1), pp.265–270

[17] Hergt, R. And Dutz, S., Magnetic Particle Hyperthermia-Biophysical Limitations of a Visionary Tumor Therapy, Journal of Magnetism and Magnetic Materials, 2007, 311 (1), pp.187-192

[18] Gonzales, M., Zeisberger, M. and Krishnan, K. M., Size-dependent Heating Rates of Iron Oxide Nanoparticles for Magnetic Fluid Hyperthermia, Journal of Magnetism and Magnetic Materials, 2009, 321 (13), pp.1947-1950

[19] Chastellain, M., Petri, A., Gupta, A., Rao, K. V., Hofmann, H., Superparamagnetic Silia-Iron Oxide Nanocomposites for Application in Hyperthermia, Advanced Engineering Materials, 2004, 6 (4), pp.235-241

[20] Fortin, J. P., Wilhelm, C., Servais, J., Menager, C., Bacri, J. C., Gazeau, F., Size-Sorted Anionic Iron Oxide Nanomagnets as Colloidal Mediators for Magnetic Hyperthermia, Journal of American Chemical Society, 2007, 129, pp.2628-2635

[21] Torres, T. E., Roca, A. G., Morales, M. P., Ibarra, A., Marquina, C., Ibarra, M. R., Goya, G. F., Magnetic Properties and Energy Absorption of $CoFe_2O_4$ Nanoparticles for Magnetic Hyperthermia, Journal of Physics-Conference Series, 2010, 200, 072101

[22] Zeisberger, M., Dutz, S., Müller, R., Hergt, R., Matoussevitch, N., Bönnemann, H., Metallic Cobalt Nanoparticles for Heating Applications, Journal of Magnetism and Magnetic Materials, 2007, 311 (1), pp. 224-227

[23] Johannsen, M., Gneveckow, U, Thiesen, B., Taymoorian, K., Cho, C. H., Waldoefner, N., Scholz, R., Jordan, A., Loening, S. A., Wust, P., Thermotherapy of Prostate Cancer Using Magnetic Nanoparticles: Feasibility, Imaging and Three-Dimensional Temperature Distribution, European Urology, 2007, 52 (6), pp. 1653-1662

[24] Van Landeghem, F. K., Maier-Hauff, K., Jordan, A., Hoffmann, K. T., Gneveckow, U., Scholz, R., Thiesen, B., Brück, W., Von Deimling, A., Post-mortem Studies in Glioblastoma Patients Treated with Thermotherapy Using Magnetic Nanoparticles, Biomaterials, 2009, 30 (1), pp. 52-57

[25] Maier-Hauff, K,, Rothe, R., Scholz, R., Gneveckow, U., Wust, P., Thiesen, B., Feussner, A., Von Deimling, A., Waldoefner, N., Felix, R., Jordan, A., Intracranial Thermotherapy Using Magnetic Nanoparticles Combined with External Beam Radiotherapy: Results of a Feasibility Study on Patients with Glioblastoma Multiforme, Journal of Neuro-oncology, 2007, 81 (1), pp.53-60.

[26] Babincov, M., Altanerova, V., Altaner, C., Bergemann, C., Babinec, P., In Vitro Analysis of Cisplatin Functionalized Magnetic Nanoparticles in Combined Cancer Chemotherapy and Electromagnetic Hyperthermia; Ieee Transactions on Nanobioscience, 2008, 7 (1), pp. 9-15

[27] Jain, T. K., Morales, M. A., Sahoo, S. K., Leslie-Pelecky, D. L., Labhasetwar, V., Iron Oxide Nanoparticles for Sustained delivery of Anticancer Agents, Molecular Pharmaceutics, 2005, 2(3), pp.194-205

[28] Gupta, A. K. and Curtis, A. S. G., Surface Modified Superparamagnetic Nanoparticles for Drug Delivery: Interaction Studies with Human Fibroblasts in Culture, Journal of Material Science: Materials in Medicine, 2004, 15 (4), pp. 493-496

[29] Yu, M. K., Jeong, Y. Y., Park, J., Park, S., Kim, J. W., Min, J. J., Kim, K., Jon, S., Drug-Loaded Superparamagnetic Iron Oxide Nanoparticles for Combined Cancer Imaging and Therapy In Vivo, Angewandte Chemie, 2008, 47 (29), pp. 5362-5365

[30] Chertok, B., Moffat, B. A., David, A. E., Yu, F., Bergemann, C., Ross, B. D. and Yang, V. C., Iron Oxide Nanoparticles as a Drug Delivery Vehicle for MRI Monitored Magnetic Targeting of Brain Tumors, Biomaterials, 2008, 29 (4), pp. 487-496

[31] Jain, T. K., Richey, J., Strand, M., Leslie-Pelecky, D. L., Flask, C. A., Labhasetwar, V., Magnetic Nanoparticles with Dual Functional Properties: Drug Delivery and Magnetic Resonance Imaging, Biomaterials, 2008, 29 (29), pp. 4012–4021

[32] Goldmann, A., Modern Ferrite Technology, Springer Publication Inc., 2006

[33] Lu, J., Ma, S., Sun, J., Xia, C., Liu, C., Wang, Z., Zhao, X., Gao, F., Gong, Q., Song, B., Shuai, X., Ai, H., Gu, Z., Manganese Ferrite Nanoparticle Micellar Nanocomposites as MRI Contrast Agent for Liver Imaging, Biomaterials, 2009, 30 (15), pp. 2919-2928

[34] Barcena, C., Sra, A. K., Chaubey, G. S., Khemtong, C., Liu, J. P. and Gao, J., Zinc Ferrite Nanoparticles as MRI Contrast Agents, Chemistry Communications, 2008, pp. 2224-2226

[35] Arulmurugan, R., Jeyadevan, B., Vaidyanathan, G. and Sendhilnathan, S., Effect of Zinc Substitution on CO-Zn and Mn-Zn Ferrite Nanoparticles Prepared by Co-precipitation, Journal of Magnetism and Magnetic Materials, 2005, 288, pp. 470-477

[36] Pollert, E., Knizek, K., Marysko, M., Kaspar, P., Vasseur, S. and Duguet, E., New Tc-tuned Magnetic Nanoparticles for Self-Controlled Hyperthermia, Journal of Magnetism and Magnetic Materials, 2007, 316 (2), pp. 122-125

[37] Maenosono, S., Suzuki, T. and Saita, S., Superparamagnetic FePt Nanoparticles as Excellent MRI Contrast Agents, of Magnetism and Magnetic Materials, 2008, 320 (9), pp. L79-L83

[38] Maenosono, S. and Saita, S., Theoretical Assessment of FePt Nanoparticles as Heating Elements for Magnetic Hyperthermia, IEEE Transactions on Magnetics, 2006, 42 (6), pp. 1638-1642

[39] Hai, N. H., Lemoine, R., Remboldt, S., Strand, M., Shield, J. E., Schmitter, D., Kraus, R. H., Espy, M., Leslie-Pelecky, D. L., Iron and Cobalt-based Magnetic Fluids Produced by Inert Gas Condensation, Journal of Magnetism and Magnetic Materials, 2005, 293 (1), pp.75-79

[40] Chakka, V. M., Altuncevahir, B., Jin, Z. Q., Li, Y. and Liu, J. P., Magnetic Nanoparticles Produced by Surfactant-asisted Ball Milling, Journal of Applied Physics, 2006, 99, pp. 08E912

[41] Wang, W., Itoh, Y., Lenggoro, I. W. And Okuyama, K., Nickel and Nickel Oxide Nanoparticles Prepared from Nickel Nitrate Hexahydrate by a Low Pressure Spray Pyrolysis, Materials Science and Engineering: B, 2004, 111 (1), pp. 69-76

[42] Chen, D. H. and He, X. R., Synthesis of Nickel Ferrite Nanoparticles by Sol-gel Method, Materials Research Bulletin, 2001, 36 (7-8), pp. 1369-1377

[43] Klug, K. L., Dravid, V. P. and Johnson D. L., Silica-encapsulated Magnetic Nanoparticles Formed by a Combined Arc-evaporation / Chemical Vapor Deposition Technique, Journal of Material Research Society, 2003, 18 (4), pp. 988-993

[44] Maaz, K., Mumtaz, A., Hasanain, S. K. and Ceylan, A., Synthesis and Magnetic Properties of Cobalt Ferrite ($CoFe_2O_4$) Nanoparticles Prepared by Wet Chemical Route, Journal of Magnetism and Magnetic Materials, 2007, 308 (2), pp. 289-295

[45] Sun, S. and Zeng, H., Size-Controlled Synthesis of Magnetite Nanoparticles, Journal of American Chemical Society, 2002, 124 (28), pp. 8204-8205

[46] Sun, S., Zeng, H., Robinson, D. B., Raoux, S., Rice, P. M., Wang, S. X. and Li, G., Monodisperse MFe_2O_4 (M=Fe, Co, Mn) Nanoparticles, Journal of American Chemical Society, 2004, 126 (1), pp. 273-279

[47] Sun, S., Anders, S., Thomson, T., Baglin, J. E. E., Toney, M. F., Hamann, H. F., Murray, C. B., Terris, B. D., Controlled Synthesis and Assembly of FePt Nanoparticles, Journal of Physical Chemistry B, 2003, 107 (23), pp. 5419-5425

[48] Chen, J. P., Lee, K. M., Sorensen, M., Klabunde, K. J., Hadjipanayis, G. C., Magnetic Propertie of Microemulsion Synthesized Cobalt Fine Particles, Journal of Applied Physics, 1994, 75 (10), pp. 5876-5878

[49] Kodama, R. H., Magnetic Nanoparticles, Journal of Magnetism and Magnetic Materials, 1999, 200 (1-3), pp. 359-372

[50] Labaye, Y., Crisan, O., Berger, L., Greneche, J. M., Coey, J. M. D., Surface Anisotropy in Ferromagnetic Nanoparticles, Journal of Applies Physics, 2002, 91 (10), pp. 8715-8717

[51] Vestal, C. R. and Zhang, Z. J., Effects of Surface Coordination Chemistry on the Magnetic Properties of $MnFe_2O_4$ Spinel Ferrite Nanoparticles, Journal of American Chemical Society, 2003, 125, pp. 9828-9833

[52] Ngo, A. T., Bonville, P. and Pileni, M. P., Nanoparticles of $CoFe_2O_4$: Synthesis and Superparamagnetic Properties, European Physical Journal B, 1999, 9 (4), pp. 583-592

[53] Tronc, E. and Jolivet, J. P., Surface Effects on Magnetically Coupled γ-Fe_2O_3 Colloids, Hyperfine Interactions, 1986, 28 (1-4), pp. 525-528

[54] Dormann, J. L., Bessais, L. and Fiorani, D., A Dynamic Study of Small Interacting Particles: Superparamagnetic Model and Spin-glass Laws, Journal of Physics C, 1988, 21 (10), pp. 2015-2034.

[55] Dormann, J. L., Cherkaoui, R., Spinu, L., Nogues, M., Lucari, F., D' Orazio, F., Fiorani, D., Garcia, A., Tronc, E., Jolivet, J. P., From Pure Superparamagnetic Regime to Glass Collective State of Magnetic Moments in γ-Fe_2O_3 Nanoparticle Assemblies, Journal of Magnetism and Magnetic Materials, 1998, 187, pp. 139-144.

[56] Fiorani, D., Dormann, J. L., Cherkaoui, R., Tronc, E., Lucari, F., D' Orazio, F., Spinu, L., Nogues, M., Garcia, A., Testa, A. M., Collective Magnetic State in Nanoparticle Systems, Journal of Magnetism and Magnetic Materials, 1999, 196-197, pp. 143-147.

[57] Sharifi, I., Shokrollahi, H. and Amiri, S., Ferrite-based Magnetic Nanofluids Used in Hyperthermia Applications, Journal of Magnetism and Magnetic Materials, 2012, 324, pp. 903-915

[58] Gupta, A. K. and Gupta, M., Synthesis and Surface Engineering of Iron Oxide Nano-particles for Biomedical Applications, Biomaterials, 2005, 26, pp. 3995-4021.

[59] Fang, C. and Zhang, M., Multifunctional Magnetic Nanoparticles for Medical Imaging Applications, Journal of Materials Chemistry, 2009, 19, pp. 6258-6266.

[60] Kumar, C.,Nanotechnologies for the Life Sciences Vol.1: Biofunctionalization of Nanomaterials, 2005, Wiley-VCH Pub.

[61] Huang, C., Hao, Y., Nyagilo, J., Dave D. P., Xu, L., Sun, X., Porous Hollow Gold Nanoparticles for Cancer SERS İmaging, Journal of Nano Research, 2010, 10, pp. 137-148

[62] Shi, W., Zeng, H., Sahoo, Y., Ohulchansky, Y., Ding, Y., Wang, Z. L., Swihart, M., Prasad, P. N., A General Approach to Binary and Ternary Hybrid Nanocrystals, Nano-letters, 2006, 6(4), pp.857-881

[63] Yu, H., Chen, M., Rice, P. M., Wang, S. X., White, R. L., Sun, S., Dumbbell-like Bifunctional Au-Fe$_3$O$_4$ Nanoparticles, Nanoletters, 2005, 5 (2), pp. 379-382.

[64] Ji, X., Shao, R., Elliot, A. M., Stafford, R., J., Esparza-Coss, E., Bankson, J. A., Liang, G., Luo, Z., Park, K., Markert, J. T., Li, C., Bifunctional Gold Nanoshells with a Super-paramagnetic Iron Oxide-Silica Core Suitable for Both MR Imaging and Photother-mal Therapy, Journal of Physical Chemistry C., 2007, 111, pp. 6245-6251.

[65] Park, H. Y., Schadt, M. J., Lim, S., Njoki, P. N., Kim, S. H., Jang, M. Y., Luo, J., Zhong, C. J., Fabrication of Magnetic Core@Shell Fe Oxide@Au Nanoparticles for Interfacial Bioactivity and Bio-separation, Langmuir, 2007, 23 (17), pp. 9050-9056.

[66] Wang, L., Bai, J., Li, Y. and Huang, Y., Multifunctional Nanoparticles Displaying Magnetization and Near-IR Absorbtion, Angewandte Chemie, 2008, 47 (13), pp. 2439-2442.

[67] Wang, C. and Irudayaraj, J., Multifunctional Magnetic-Optical Nanoparticles Probes for Simultaneous Detection, Separation and Thermal Ablation of Multiple Pathogens, Small, 2010, 6 (2), pp. 283-289.

[68] Chen, W., Xu, N., Xu, L., Wang, L., Li, Z., Ma, W., Zhu, Y., Xu, C., Kotov, N. A., Mul-tifunctional Magnetoplasmonic Nanoparticle Assemblies for Cancer Therapy and Di-agnostics (Theranostics), Macromolecular Rapid Communications, 2010, 31 (2), pp. 228-236.

[69] Park, H., Yang, J., Seo, S., Kim, K., Suh, J., Kim, D., Haam, S., Yoo, K. H., Multifunctional Nanoparticles for Photothermal Controlled Drug Delivery and Magnetic Resonance Imaging Enhancement, Small, 2008, 4 (2), pp. 192-196.

[70] Ulman, A., Formation and Structure of Self-Assembled Monolayers, Chemical Reviews, 1996, 96, pp. 1533-1554

[71] Stöber, W., Fink, A. and Bohn, E., Controlled Growth of Monodispersed Silica Spheres in the Micron Size Range, Journal of Colloid and Interface Science, 1968, 26 (1), pp. 62-69.

[72] Deng, Y. H., Wang, C. C., Hu, J. H., Yang, W. L., Fu, S. K., Investigation of Formation of Silica-Coated Magnetite Nanoparticles via Sol-gel Approach, Colloids and Surfaces A: Physicochemical and Engineering Aspects, 2005, 262 (1-3), pp. 87-93.

[73] Santra, S., Tapec, R., Theodoropoulou, N., Dobson, J., Hebard, A., Tan, W., Synthesis and Characterization of Silica-Coated Iron Oxide Nanoparticles in Microemulsion: The Effect of Nonionic Surfactants, Langmuir, 2001, 17 (10), pp. 2900-2906.

[74] Yi, D. K., Lee, S. S., Papaefthymiou, G. C. and Ying, J. Y., Nanoparticle Architectures Templated by SiO_2/Fe_2O_3 Nanocomposites, Chemistry of Materials, 2006, 18, pp. 614-619.

[75] Yoon, T. J., Yu, K. M., Kim, E., Kim, J. S., Kim, B. G., Yun, S. H., Sohn, B. H., Cho, M. H., Lee, J. K., Park, S. B., Specific Targeting, Cell Sorting and Bioimaging with Smart Magnetic Silica Core-Shell Nanomaterials, Small, 2006, 2 (2), pp. 209-215.

[76] Yi, D. K., Selvan, S. T., Lee, S. S., Papaefthymiou, G. C., Kundaliya, D., Ying, J. Y., Silica Coated Nanocomposites of Magnetic Nanoparticles and Quantum Dots, Journal of American Chemical Society, 2005, 127, pp. 4990-4991.

[77] Chang, Y. and Chen, D, Preparation and Adsorbtion Properties of Monodisperse Chitosan-Bound Fe_3O_4 Magnetic Nanoparticles for Removal of Cu(II) ions, Journal of Colloid and Interface Science, 2005, 283 (2), pp. 446-451.

[78] Zhi, J., Wang, Y., Lu, Y., Ma, J., Luo, G., In Situ Preparation of Magnetic Chitosan/ Fe_3O_4 Composite Nanoparticles in Tiny Pools of Water-in-Oil Microemulsion, Reactive and Functional Polymers, 2006, 66 (12), pp. 1552-1558.

[79] Kim, D. H., Kim, K. N., Kim, K. M., Lee, Y. K., Targeting to Carcinoma Cells with Chitosan- and Starch-Coated Magnetic Nanoparticles for Magnetic Hyperthermia, Journal of Biomedical Materials Research A, 2009, 88A (1), pp. 1-11.

[80] Ge. Y., Zhang, Y., He, S., Nie, F., Teng, G., Gu, N., Fluorescence Modified Chitosan-Coated Magnetic Nanoparticles for High-Efficient Cellular Imaging, Nanoscale Research Letters, 2009, 4, pp. 287-295.

[81] Crepon, B., Chytry, J. J., Kopecek, R., Enzymatic Degradation and Immunogenic Properties of Derivatized Dextrans, Biomaterials, 1991,12(6), pp. 550–554

[82] Jordan, A., Scholz, R., Wust, P., Schirra, H., Schiestel, T., Schmidt, H., Felix, R., Endocytosis of Dextran and Silica-Coated Magnetite Nanoparticles and the Effect of Intercellular Hyperthermia on Human Mammary Carcinoma Cells in Vitro, Journal of Magnetism and Magnetic Materials, 1999, 194 (1-3), pp. 185-196.

[83] Moore, A., Basilion, J., Chiocca, E. A., Weissleder, R., Measuring Transferrin Receptor Gene Expression by NMR Imaging, Biochimica et Biophysica Acta, 1998, 1402, pp. 239–249.

[84] Weissleder, R., Cheng, H. C., Bogdanova, A., Bogdanov, A., Magnetically Labelled Cells Can Be Detected by MR Imaging. Magnetic Resonance Imaging, 1997, 7, pp. 258–263.

[85] Bautista, M. C., Miguel, O. B., Morales, M. P., Serna, C. J., Verdaguer, S. V., Surface Characterization of Dextran-coated Iron Oxide Nanoparticles Prepared by Laser Pyrolysis and Coprecipitation, Journal of Magnetism and Magnetic Materials, 2005, 293 (1), pp. 20-27.

[86] Jarrett, B. R., Frendo, M., Vogan, J. and Louie, A. Y., Size-controlled Synthesis of Dextran Coated Iron Oxide Nanoparticles for Magnetic Resonance Imaging, Nanotechnology, 2007, 18, pp. 035603.

[87] Moore, A., Weissleder, R. and Bogdanov, A., Uptake of Dextran-Coated Monocrystalline Iron Oxide in Tumor Cells and Macrophages, Journal of Magnetic Resonance Imaging, 1997, 7 (6), pp. 1140-1145

[88] Berry, C. C., Possible Exploitation of Magnetic Nanoparticle-Cell Interaction for Biomedical Applications, Journal of Materials Chemistry, 2005, 15, pp. 543-547.

[89] Zhang, Y., Kohler, N. and Zhang, M., Surface Modification of Superparamagnetic Magnetite Nanoparticles and Their Intracellular Uptake, Biomaterials, 2002, 23 (7), pp.1553–1561.

[90] Yamazaki, M. and Ito, M., Deformation and Instability of Membrane Structure of Phospholipid Vesicles Caused by Osmophobic Association: Mechanical Stress Model for the Mechanism of Poly(ethylene glycol)-induced Membrane Fusion, Biochemistry, 1990, 29 (5), pp. 1309–1314.

[91] Suzuki, M., Shinkai, M., Kamihira, M., Kobayashi, T., Preparation and Characteristics of Magnetite-Labeled Antibody with the Use of Poly(ethylene glycol) Derivatives. Biotechnology and Applied Biochemistry, 1995, 21, pp. 335–345.

[92] Lee, J., Isobe, T. and Senna, M., Preparation of Ultrafine Fe_3O_4 Particles by Precipitation in the Presence of PVA at high pH., Journal of Colloid and Interface Science, 1996, 177 (2), pp. 490-494.

[93] Sahoo, S. K., Panyam, J., Prabha, S. and Labhasetwar, V., Residual Polyvinyl Alcohol Associated with Poly (D,L-Lactide-Co-Glycolide) Nanoparticles Affects Their Physi-

cal Properties and Cellular Uptake, Journal of Controlled Release, 2002, 82 (1), pp. 105-114.

[94] Petri-Fink, A., Chastellain, M., Juillerat-Jenneret, L., Ferrari, A.,Hofmann, H., Development of Functionalized Superparamagnetic Iron Oxide Nanoparticles for Interaction with Human Cancer Cells, Biomaterials, 2005, 26, pp. 2685–2694.

[95] Kircheis, R., Wightman, L. and Wagner, E., Design and Gene Delivery Activity of Modified Polyethyleneimine, Advanced Drug Delivery Reviews, 2001, 53 (3), pp. 341-358.

[96] Kievit, F. M., Veiseh, O., Bhattarai, N., Fang, C., Gunni J. W., Lee, D., Ellenbogen, R. G., Olson, J. M., Zhang, M., PEI–PEG–Chitosan-Copolymer-Coated Iron Oxide Nanoparticles for Safe Gene Delivery: Synthesis, Complexation, and Transfection, Advanced Functional Materials, 2009, 19 (14), pp. 2244-2251.

Permissions

The contributors of this book come from diverse backgrounds, making this book a truly international effort. This book will bring forth new frontiers with its revolutionizing research information and detailed analysis of the nascent developments around the world.

We would like to thank Dr. Mahmood Aliofkhazraei, for lending his expertise to make the book truly unique. He has played a crucial role in the development of this book. Without his invaluable contribution this book wouldn't have been possible. He has made vital efforts to compile up to date information on the varied aspects of this subject to make this book a valuable addition to the collection of many professionals and students.

This book was conceptualized with the vision of imparting up-to-date information and advanced data in this field. To ensure the same, a matchless editorial board was set up. Every individual on the board went through rigorous rounds of assessment to prove their worth. After which they invested a large part of their time researching and compiling the most relevant data for our readers. Conferences and sessions were held from time to time between the editorial board and the contributing authors to present the data in the most comprehensible form. The editorial team has worked tirelessly to provide valuable and valid information to help people across the globe.

Every chapter published in this book has been scrutinized by our experts. Their significance has been extensively debated. The topics covered herein carry significant findings which will fuel the growth of the discipline. They may even be implemented as practical applications or may be referred to as a beginning point for another development. Chapters in this book were first published by InTech; hereby published with permission under the Creative Commons Attribution License or equivalent.

The editorial board has been involved in producing this book since its inception. They have spent rigorous hours researching and exploring the diverse topics which have resulted in the successful publishing of this book. They have passed on their knowledge of decades through this book. To expedite this challenging task, the publisher supported the team at every step. A small team of assistant editors was also appointed to further simplify the editing procedure and attain best results for the readers.

Our editorial team has been hand-picked from every corner of the world. Their multi-ethnicity adds dynamic inputs to the discussions which result in innovative outcomes. These outcomes are then further discussed with the researchers and contributors who give their valuable feedback and opinion regarding the same. The feedback is then collaborated with the researches and they are edited in a comprehensive manner to aid the understanding of the subject.

Apart from the editorial board, the designing team has also invested a significant amount of their time in understanding the subject and creating the most relevant covers. They scrutinized every image to scout for the most suitable representation of the subject and create an appropriate cover for the book.

The publishing team has been involved in this book since its early stages. They were actively engaged in every process, be it collecting the data, connecting with the contributors or procuring relevant information. The team has been an ardent support to the editorial, designing and production team. Their endless efforts to recruit the best for this project, has resulted in the accomplishment of this book. They are a veteran in the field of academics and their pool of knowledge is as vast as their experience in printing. Their expertise and guidance has proved useful at every step. Their uncompromising quality standards have made this book an exceptional effort. Their encouragement from time to time has been an inspiration for everyone.

The publisher and the editorial board hope that this book will prove to be a valuable piece of knowledge for researchers, students, practitioners and scholars across the globe.

List of Contributors

Alina Brindusa Petre, Claudia Mihaela Hristodor, Aurel Pui and Diana Tanasa
"Al.I.Cuza" University of Iasi, Iasi, Romania

Narcisa Vrinceanu
"Al.I.Cuza" University of Iasi, Iasi, Romania
"L.Blaga" University of Sibiu, Romania

Diana Coman
"L.Blaga" University of Sibiu, Romania

Eveline Popovici
"Al.I.Cuza" University of Iasi, Faculty of Chemistry, Department of Materials Chemistry, Romania

R. M. Miranda
Mechanical and Industrial Engineering Department, Sciences and Technology Faculty, Nova University of Lisbon, Caparica, Portugal

J. Gandra and P. Vilaça
Mechanical Engineering Department, Lisbon Technical University, Av. Rovisco Pais, Lisboa, Portugal

Jim C.E. Odekerken, Tim J.M. Welting, Jacobus J.C. Arts, Geert H.I.M. Walenkamp and Pieter J. Emans
Department of Orthopaedic Surgery, Research school CAPHRI, Maastricht University Medical Centre, the Netherlands

R. Abdel-Karim
Department of Metallurgy, Faculty of Engineering, Cairo University, Giza, Egypt

A. F. Waheed
Department of Metallurgy, Nuclear Research Center, Cairo, Anshas, Egypt

Jun Liang
State Key Laboratory of Solid Lubrication, Lanzhou Institute of Chemical Physics, Chinese Academy of Sciences, Lanzhou, PR China

Qingbiao Li
State Key Laboratory of Solid Lubrication, Lanzhou Institute of Chemical Physics, Chinese Academy of Sciences, Lanzhou, PR China
School of Science, Lanzhou University of Technology, Lanzhou, PR China

Qing Wang
School of Science, Lanzhou University of Technology, Lanzhou, PR China

Eric M. Garcia
Federal University of São João Del Rei- unit of Sete Lagoas, Minas Gerais, Sete Lagoas, Brazil

Vanessa F.C. Lins and Tulio Matencio
Federal University of Minas Gerais, Minas Gerais, Belo Horizonte, Brazil

Malin Liu
Institute of Nuclear and New Energy Technology, Tsinghua Univerisity, Beijing, China

Paulo S. da Silva and Almir Spinelli
Chemistry Department, Federal University of Santa Catarina, Florianópolis, SC, Brazil

Jose M. Maciel, Karen Wohnrath and Jarem R. Garcia
Chemistry Department, State University of Ponta Grossa, Ponta Grossa, PR, Brazil

Evrim Umut
Hacettepe University, Department of Physics Engineering, Ankara, Turkey

Printed in the USA
CPSIA information can be obtained
at www.ICGtesting.com
JSHW011423221024
72173JS00004B/649

9 781632 381804